TEACHER GUIDE
3rd–8th Grade

Includes Student | Weekly Lesson Schedule
Student Worksheets
Final Exam

God's Design: Chemistry & Ecology

Master Books Creative Team:
Authors: Richard and Debbie Lawrence
Editor: Craig Froman
Design: Terry White
Cover Design: Diana Bogardus
Copy Editors: Judy Lewis, Willow Meek
Curriculum Review: Kristin Pratt, Laura Welch, Diana Bogardus

First printing: July 2018
Fourth printing: March 2022

Copyright © 2018 by Debbie and Richard Lawrence and Master Books®. All rights reserved. No part of this book may be reproduced, copied, broadcast, stored, or shared in any form whatsoever without written permission from the publisher, except in the case of brief quotations in articles and reviews. For information write:

Master Books®, P.O. Box 726, Green Forest, AR 72638

Master Books® is a division of the New Leaf Publishing Group, Inc.

ISBN: 978-1-68344-126-7
ISBN: 978-1-61458-650-0 (digital)

Scripture taken from the New King James Version. Copyright © 1982 by Thomas Nelson, Inc. Used by permission. All rights reserved.

Printed in the United States of America

Please visit our website for other great titles:
www.masterbooks.com

Permission is granted for copies of reproducible pages from this text to be made for use with immediate family members living in the same household. However, no part of this book may be reproduced, copied, broadcast, stored, or shared in any form beyond this use. Permission for any other use of the material must be requested by email from the publisher at info@nlpg.com.

The God's Design Science series is based on a biblical worldview and reveals how science supports the biblical account of creation. **Richard and Debbie Lawrence** have a long history of enjoying science. They have both worked as electrical engineers and now Debbie teaches chemistry and physics at a homeschool co-op. While homeschooling their children for 16 years, there was almost always a science experiment going on in the kitchen. Today that tradition is being continued with the next generation as the grandkids enjoy Grandma Science Day once a week.

> Your reputation as a publisher is stellar. It is a blessing knowing anything I purchase from you is going to be worth every penny!
> —Cheri ★★★★★

> Last year we found Master Books and it has made a HUGE difference.
> —Melanie ★★★★★

> We love Master Books and the way it's set up for easy planning!
> —Melissa ★★★★★

> You have done a great job. MASTER BOOKS ROCKS!
> —Stephanie ★★★★★

> Physically high-quality, Biblically faithful, and well-written.
> —Danika ★★★★★

> Best books ever. Their illustrations are captivating and content amazing!
> —Kathy ★★★★★

Affordable
Flexible
Faith Building

Table of Contents

Welcome to God's Design 5

Daily Suggested Schedule 15

Properties of Matter Worksheets 23

Properties of Ecosystems Worksheets 121

Properties of Atoms & Molecules Worksheets 239

Properties of Matter Quizzes and Final Exam 337

Properties of Ecosystems Quizzes and Final Exam 355

Properties of Atoms & Molecules Quizzes and Final Exam 375

Answer Keys

    Properties of Matter Worksheets 397

    Properties of Ecosystems Worksheets 409

    Properties of Atoms & Molecules Worksheets 421

    Properties of Matter Quizzes 435

    Properties of Ecosystems Quizzes 439

    Properties of Atoms & Molecules Quizzes 445

    Properties of Matter Final Exam 451

    Properties of Ecosystems Final Exam 453

    Properties of Atoms & Molecules Final Exam 455

Appendices 457

Teacher Note: As with any science course that includes experiments, what is created can be potentially hazardous if not handled properly. Make sure to follow all instructions very carefully:

- wear proper safety equipment when needed, including safety goggles/glasses
- keep small children away from where the labs are conducted
- wash hands, surfaces, and equipment properly after each experiment
- and make sure clothing and other household surfaces are protected from staining.

Welcome to GOD'S DESIGN®

CHEMISTRY & ECOLOGY

God's Design for Chemistry & Ecology has been designed for use in teaching chemistry and ecology to elementary and middle school students. It is divided into three sections: *Properties of Matter*, *Properties of Atoms & Molecules*, and *Properties of Ecosystems*. Each has 35 lessons including a final project that ties all of the lessons together.

In addition to the lessons, special features in each include biographical information on interesting people as well as fun facts to make the subject more fun.

Although this is a complete curriculum, the information included here is just a beginning, so please feel free to add to each lesson as you see fit. A resource guide is included in the appendices to help you find additional information and resources. A list of supplies needed is included at the beginning of each lesson, while a master list of all supplies needed for the course can be found in the appendices.

Answer keys for all review questions, worksheets, quizzes, and the final exam are included here.

If you wish to get through all of *Chemistry & Ecology* in one year, you should plan on covering approximately four lessons per week. The time required for each lesson varies depending on how much additional information you want to include, but you can plan on about 45 minutes per lesson.

If you wish to cover the material in more depth, you may add additional information and take a longer period of time to cover all the material or you could choose to do only one or two of the sections as a unit study.

Why Teach Chemistry & Ecology?

Maybe you hate science or you just hate teaching it. Maybe you love science but don't quite know how to teach it to your children. Maybe science just doesn't seem as important as some of those other subjects you need to teach. Maybe you need a little motivation. If any of these descriptions fit you, then please consider the following.

God is the Master Creator of everything. His handiwork is all around us. Our great Creator put in place all of the laws of physics, biology, and chemistry. These laws were put here for us to see His wisdom and power. In science, we see the hand of God at work more than in any other subject. Romans 1:20 says, "For since the creation of the world His invisible attributes are clearly seen, being understood by the things that are made, even His eternal power and Godhead, so that they [men] are without excuse." We need to help our children see God as Creator of the world around them so they will be able to recognize God and follow Him.

The study of chemistry helps us understand and appreciate the amazing way everything God created works together. The study of atoms and molecules and how different substances react with each other reveals an amazing design, even at the smallest level of life. Understanding the carbon, nitrogen, and water cycles helps our children see that God has a plan to keep everything working together. Learning about ecosystems reveals God's genius in nature.

It's fun to teach chemistry and ecology! It's interesting too. The elements of chemistry are all around us. Children naturally like to combine things to see what will happen. You just need to direct their curiosity.

Finally, teaching chemistry is easy. You won't have to try to find strange materials for experiments or do dangerous things to learn about chemistry. Chemistry is as close as your kitchen or your own body, and ecosystems are just outside your door.

How Do I Teach Science?

In order to teach any subject you need to understand how people learn. People learn in different ways. Most people, and children in particular, have a dominant or preferred learning style in which they absorb and retain information more easily.

If a student's dominant style is:
Auditory
He or she needs not only to hear the information, but to hear him or herself say it. This child needs oral presentation as well as oral drill and repetition.
Visual
The student needs things he or she can see. This child responds well to flashcards, pictures, charts, models, etc.
Kinesthetic
He or she needs active participation. This child remembers best through games, hands-on activities, experiments, and field trips.

Also, some people are more relational while others are more analytical. The relational student needs to know why this subject is important, and how it will affect him or her personally. The analytical student, however, wants just the facts.

If you are trying to teach more than one student, you will probably have to deal with more than one learning style. Therefore, you need to present your lessons in several different ways so that each student can grasp and retain the information.

Grades 3–8

The first part of each lesson should be completed by all upper elementary and junior high students. This is the main part of the lesson containing a reading section, a hands-on activity that reinforces the ideas in the reading section (blue box), and a review section that provides review questions and application questions.

Grades 6–8

In addition, for middle school/junior high age students, we provide a "Challenge" section that contains more challenging material as well as additional activities and projects for older students (green box).

We have included periodic biographies to help your students appreciate the great men and women who have gone before us in the field of science.

We suggest a threefold approach to each lesson:

Introduce the topic

We give a brief description of the facts. Frequently you will want to add more information than the essentials given in this book. In addition to reading this section aloud (or having older children read it on their own), you may wish to do one or more of the following:

- Read a related book with your students.
- Write things down to help your visual learners.
- Give some history of the subject. We provide some historical sketches to help you, but you may want to add more.
- Ask questions to get your students thinking about the subject.

Make observations and do experiments

- Hands-on projects are suggested for each lesson. This part of each lesson may require help from the teacher.
- Have your students perform the activity by themselves whenever possible.

Review

- The "What did we learn?" section has review questions.
- The "Taking it further" section encourages students to:
  - Draw conclusions
  - Make applications of what was learned
  - Add extended information to what was covered in the lesson
- The "FUN FACT" section adds fun or interesting information.

By teaching all three parts of the lesson, you will be presenting the material in a way that children with any learning style can both relate to and remember.

Also, this approach relates directly to the scientific method and will help your students think more scientifically. The *scientific method* is just a way to examine a subject logically and learn from it. Briefly, the steps of the scientific method are:

1. Learn about a topic.
2. Ask a question.
3. Make a hypothesis (a good guess).
4. Design an experiment to test your hypothesis.
5. Observe the experiment and collect data.
6. Draw conclusions. (Does the data support your hypothesis?)

Note: It's okay to have a "wrong hypothesis." That's how we learn. Be sure to help your students understand why they sometimes get a different result than expected.

Our lessons will help your students begin to approach problems in a logical, scientific way.

How Do I Teach Creation vs. Evolution?

We are constantly bombarded by evolutionary ideas about the earth in books, movies, museums, and even commercials. These raise many questions: What is the big bang? How old is the earth? Do fossils show evolution to be true? Was there really a worldwide flood? When did dinosaurs live? Was there an ice age? How can we teach our children the truth about the origins of the earth? The Bible answers these questions, and this book accepts the historical accuracy of the Bible as written. We believe this is the only way we can teach our children to trust that everything God says is true.

There are five common views of the origins of life and the age of the earth:

Historical biblical account	Progressive creation	Gap theory	Theistic evolution	Naturalistic evolution
Each day of creation in Genesis is a normal day of about 24 hours in length, in which God created everything that exists. The earth is only thousands of years old, as determined by the genealogies in the Bible.	The idea that God created various creatures to replace other creatures that died out over millions of years. Each of the days in Genesis represents a long period of time (day-age view), and the earth is billions of years old.	The idea that there was a long, long time between what happened in Genesis 1:1 and what happened in Genesis 1:2. During this time, the "fossil record" was supposed to have formed, and millions of years of earth history supposedly passed.	The idea that God used the process of evolution over millions of years (involving struggle and death) to bring about what we see today.	The view that there is no God, and evolution of all life forms happened by purely naturalistic processes over billions of years.

Any theory that tries to combine the evolutionary time frame with creation presupposes that death entered the world before Adam sinned, which contradicts what God has said in His Word. The view that the earth (and its "fossil record") is hundreds of millions of years old damages the gospel message. God's completed creation was "very good" at the end of the sixth day (Genesis 1:31). Death entered this perfect paradise *after* Adam disobeyed God's command. It was the punishment for Adam's sin (Genesis 2:16–17, 3:19; Romans 5:12–19). Thorns appeared when God cursed the ground because of Adam's sin (Genesis 3:18).

The first animal death occurred when God killed at least one animal, shedding its blood, to make clothes for Adam and Eve (Genesis 3:21). If the earth's "fossil record" (filled with death, disease, and thorns) formed over millions of years before Adam appeared (and before he sinned), then death no longer would be the penalty for sin. Death, the "last enemy" (1 Corinthians 15:26), diseases (such as cancer), and thorns would instead be part of the original creation that God labeled "very good." No, it is clear that the "fossil record" formed some time *after* Adam sinned—not many millions of years before. Most fossils were formed as a result of the worldwide Genesis Flood.

When viewed from a biblical perspective, the scientific evidence clearly supports a recent creation by God, and not naturalistic evolution and millions of years. The volume of evidence supporting the biblical creation account is substantial and cannot be adequately covered in this book. If you would like more information on this topic, please see the resource guide in the appendices. To help get you started, just a few examples of evidence supporting biblical creation are given on the following pages.

Evolutionary Myth: Life evolved from non-life when chemicals randomly combined together to produce amino acids and then proteins that produced living cells.

The Truth: The chemical requirements for DNA and proteins to line up just right to create life could not have happened through purely natural processes. The process of converting DNA information into proteins requires at least 75 different protein molecules. But each and every one of these 75 proteins must be synthesized in the first place by the process in which they themselves are involved. How could the process begin without the presence of all the necessary proteins? Could all 75 proteins have arisen by chance in just the right place at just the right time? Dr. Gary Parker says this is like the chicken and the egg problem. The obvious conclusion is that both the DNA and proteins must have been functional from the beginning, otherwise life could not exist. The best explanation for the existence of these proteins and DNA is that God created them.

Gary Parker, *Creation: Facts of Life* (Master Books, 2006), pp. 20–43.

Evolutionary Myth: Stanley Miller created life in a test tube, thus demonstrating that the early earth had the conditions necessary for life to begin.

The Truth: Although Miller was able to create amino acids from raw chemicals in his famous experiment, he did not create anything close to life or even the ingredients of life. There are four main problems with Miller's experiment. First, he left out oxygen because he knew that oxygen corrodes and destroys amino acids very quickly. However, rocks found in every layer of the earth indicate that oxygen has always been a part of the earth's atmosphere. Second, Miller included ammonia gas and methane gas. Ammonia gas would not have been present in any large quantities because it would have been dissolved in the oceans. And there is no indication in any of the rock layers that methane has ever been a part of the earth's atmosphere. Third, Miller used a spark of electricity to cause the amino acids to form, simulating lightning. However, this spark more quickly destroyed the amino acids than built them up, so to keep the amino acids from being destroyed, Miller used specially designed equipment to siphon off the amino acids before they could be destroyed. This is not what would have happened in nature. And finally, although Miller did produce amino acids, they were not the kinds of amino acids that are needed for life as we know it. Most of the acids were ones that actually break down proteins, not build them up.

Mike Riddle, "Can Natural Processes Explain the Origin of Life?" in *The New Answers Book 2*, Ken Ham, ed. (Master Books, 2008).

Evolutionary Myth: Living creatures are just a collection of chemicals.

The Truth: It is true that cells are made of specific chemicals. However, a dead animal is made of the same chemicals as it was when it was living, but it cannot become alive again. What makes the chemicals into a living creature is the result of the organization of the substances, not just the substances themselves. Dr. Parker again uses an example. An airplane is made up of millions of non-flying parts; however, it can fly because of the design and organization of those parts. Similarly, plants and animals are alive because God created the chemicals in a specific way for them to be able to live. A collection of all the right parts is not life.

Evolutionary Myth: Chemical evidence points to an earth that is billions of years old.

The Truth: Much of the chemical evidence actually points to a young earth. For example, radioactive decay in the earth's crust produces helium atoms that rise to the surface and enter the atmosphere. Assuming that the rate of helium production has always been constant (an evolutionary assumption), the maximum age for the atmosphere could only be 2 million years.[1] This is much younger than the 4+ billion years claimed by evolutionists. And there are many ideas that could explain the presence of helium that would indicate a much younger age than 2 million years. Similarly, salt accumulates in the ocean over time. Evolutionists claim that life evolved in a salty ocean 3–4 billion years ago. If this were true and the salt has continued to accumulate over billions of years, the ocean would be too salty for anything to live in by now. Using the most conservative possible values (those that would give the oldest possible age for the oceans), scientists have calculated that the ocean must be less than 62 million years. That number is based on the assumption that nothing has affected the rate at which the salt is accumulating. However, the Genesis Flood would have drastically altered the amount of salt in the ocean, dissolving much sodium from land rocks.[2] Thus, the chemical evidence does not support an earth that is billions of years old.

[1] Don DeYoung, *Thousands...not billions* (Master Books, 2005).
[2] John D. Morris, *The Young Earth* (Master Books, 2007), pp. 83–87.

Despite the claims of many scientists, if you examine the evidence objectively, it is obvious that evolution and millions of years have not been proven. You can be confident that if you teach that what the Bible says is true, you won't go wrong. Instill in your student a confidence in the truth of the Bible in all areas. If scientific thought seems to contradict the Bible, realize that scientists often make mistakes, but God does not lie. At one time scientists believed that the earth was the center of the universe, that living things could spring from non-living things, and that blood-letting was good for the body. All of these were believed to be scientific facts but have since been disproved, but the Word of God remains true. If we use modern "science" to interpret the Bible, what will happen to our faith in God's Word when scientists change their theories yet again?

Integrating the Seven C's

The Seven C's is a framework in which all of history, and the future to come, can be placed. As we go through our daily routines we may not understand how the details of life connect with the truth that we find in the Bible. This is also the case for students. When discussing the importance of the Bible you may find yourself telling students that the Bible is relevant in everyday activities. But how do we help the younger generation see that? The Seven C's are intended to help.

The Seven C's can be used to develop a biblical worldview in students, young or old. Much more than entertaining stories and religious teachings, the Bible has real connections to our everyday life. It may be hard, at first, to see how many connections there are, but with practice, the daily relevance of God's Word will come alive. Let's look at the Seven C's of History and how each can be connected to what the students are learning.

Creation

God perfectly created the heavens, the earth, and all that is in them in six normal-length days around 6,000 years ago.

This teaching is foundational to a biblical worldview and can be put into the context of any subject. In science, the amazing design that we see in nature—whether in the veins of a leaf or the complexity of your hand—is all the handiwork of God. Virtually all of the lessons in *God's Design for Science* can be related to God's creation of the heavens and earth.

Other contexts include:

Natural laws—any discussion of a law of nature naturally leads to God's creative power.

DNA and information—the information in every living thing was created by God's supreme intelligence.

Mathematics—the laws of mathematics reflect the order of the Creator.

Biological diversity—the distinct kinds of animals that we see were created during the Creation Week, not as products of evolution.

Art—the creativity of man is demonstrated through various art forms.

History—all time scales can be compared to the biblical time scale extending back about 6,000 years.

Ecology—God has called mankind to act as stewards over His creation.

Corruption

After God completed His perfect creation, Adam disobeyed God by eating the forbidden fruit. As a result, sin and death entered the world, and the world has been in decay since that time. This point is evident throughout the world that we live in. The struggle for survival in animals, the death of loved ones, and the violence all around us are all examples of the corrupting influence of sin.

Other contexts include:

Genetics—the mutations that lead to diseases, cancer, and variation within populations are the result of corruption.

Biological relationships—predators and parasites result from corruption.

History—wars and struggles between mankind, exemplified in the account of Cain and Abel, are a result of sin.

Catastrophe

God was grieved by the wickedness of mankind and judged this wickedness with a global flood. The Flood covered the entire surface of the earth and killed all air-breathing creatures that were not aboard the Ark. The eight people and the animals aboard the Ark replenished the earth after God delivered them from the catastrophe.

The catastrophe described in the Bible would naturally leave behind much evidence. The studies of geology and of the biological diversity of animals on the planet are two of the most obvious applications of this event. Much of scientific understanding is based on how a scientist views the events of the Genesis Flood.

Other contexts include:

Biological diversity—all of the birds, mammals, and other air-breathing animals have populated the earth from the original kinds which left the Ark.

Geology—the layers of sedimentary rock seen in roadcuts, canyons, and other geologic features are testaments to the global Flood.

Geography—features like mountains, valleys, and plains were formed as the floodwaters receded.

Physics—rainbows are a perennial sign of God's faithfulness and His pledge to never flood the entire earth again.

Fossils—most fossils are a result of the Flood rapidly burying plants and animals.

Plate tectonics—the rapid movement of the earth's plates likely accompanied the Flood.

Global warming/Ice Age—both of these items are likely a result of the activity of the Flood. The warming we are experiencing today has been present since the peak of the Ice Age (with variations over time).

Confusion

God commanded Noah and his descendants to spread across the earth. The refusal to obey this command and the building of the tower at Babel caused God to judge this sin. The common language of the people was confused and they spread across the globe as groups with a common language. All people are truly of "one blood" as descendants of Noah and, originally, Adam.

The confusion of the languages led people to scatter across the globe. As people settled in new areas, the traits they carried with them became concentrated in those populations. Traits like dark skin were beneficial in the tropics while other traits benefited populations in northern climates, and distinct people groups, not races, developed.

Other contexts include:

Genetics—the study of human DNA has shown that there is little difference in the genetic makeup of the so-called "races."

Languages—there are about seventy language groups from which all modern languages have developed.

Archaeology—the presence of common building structures, like pyramids, around the world confirms the biblical account.

Literature—recorded and oral records tell of similar events relating to the Flood and the dispersion at Babel.

Christ

God did not leave mankind without a way to be redeemed from its sinful state. The Law was given to Moses to show how far away man is from God's standard of perfection. Rather than the sacrifices, which only covered sins, people needed a Savior to take away their sin. This was accomplished when Jesus Christ came to earth to live a perfect life and, by that obedience, was able to be the sacrifice to satisfy God's wrath for all who believe.

The deity of Christ and the amazing plan that was set forth before the foundation of the earth is the core of Christian doctrine. The earthly life of Jesus was the fulfillment of many prophecies and confirms the truthfulness of the Bible. His miracles and presence in human form demonstrate that God is both intimately concerned with His creation and able to control it in an absolute way.

Other contexts include:

Psychology—popular secular psychology teaches of the inherent goodness of man, but Christ has lived the only perfect life. Mankind needs a Savior to redeem it from its unrighteousness.

Biology—Christ's virgin birth demonstrates God's sovereignty over nature.

Physics—turning the water into wine and feeding the five thousand demonstrate Christ's deity and His sovereignty over nature.

History—time is marked (in the western world) based on the birth of Christ despite current efforts to change the meaning.

Art—much art is based on the life of Christ and many of the masters are known for these depictions, whether on canvas or in music.

Cross

Because God is perfectly just and holy, He must punish sin. The sinless life of Jesus Christ was offered as a substitutionary sacrifice for all of those who will repent and put their faith in the Savior. After His death on the Cross, He defeated death by rising on the third day and is now seated at the right hand of God.

The events surrounding the crucifixion and resurrection have a most significant place in the life of Christians. Though there is no way to scientifically prove the resurrection, there is likewise no way to prove the stories of evolutionary history. These are matters of faith founded in the truth of God's Word and His character. The eyewitness testimony of over 500 people and the written Word of God provide the basis for our belief.

Other contexts include:

Biology—the biological details of the crucifixion can be studied alongside the anatomy of the human body.

History—the use of crucifixion as a method of punishment was short-lived in historical terms and not known at the time it was prophesied.

Art—the crucifixion and resurrection have inspired many wonderful works of art.

Consummation

God, in His great mercy, has promised that He will restore the earth to its original state—a world without death, suffering, war, and disease. The corruption introduced by Adam's sin will be removed. Those who have repented and put their trust in the completed work of Christ on the Cross will experience life in this new heaven and earth. We will be able to enjoy and worship God forever in a perfect place.

This future event is a little more difficult to connect with academic subjects. However, the hope of a life in God's presence and in the absence of sin can be inserted in discussions of human conflict, disease, suffering, and sin in general.

Other contexts include:

History—in discussions of war or human conflict the coming age offers hope.

Biology—the violent struggle for life seen in the predator-prey relationships will no longer taint the earth.

Medicine—while we struggle to find cures for diseases and alleviate the suffering of those enduring the effects of the Curse, we ultimately place our hope in the healing that will come in the eternal state.

The preceding examples are given to provide ideas for integrating the Seven C's of History into a broad range of curriculum activities. The first seven lessons of this curriculum cover the Seven C's and will establish a solid understanding of the true history, and future, of the universe. Full lesson plans, activities, and student resources are provided in AiG's *Answers for Kids* curriculum set if you wish to focus on the Seven C's.

Even if you use other curricula, you can still incorporate the Seven C's teaching into those. Using this approach will help students make firm connections between biblical events and every aspect of the world around them, and they will begin to develop a truly biblical worldview and not just add pieces of the Bible to what they learn in "the real world."

First Semester Suggested Daily Schedule

Date	Day	Assignment	Due Date	✓	Grade
		First Semester-First Quarter			
Week 1	Day 1	Properties of Matter Unit 1: Experimental Science Read Lesson 1: Introduction to Experimental Science Pages 14–16 • *God's Design: Chemistry & Ecology* • (GDCE) Complete Worksheet • Pages 25–26 • *Teacher Guide* • (TG)			
	Day 2	Read Lesson 2: The Scientific Method • Pages 17–20 • (GDCE) Complete Worksheet • Pages 27–29 • (TG)			
	Day 3	Read Lesson 3: Tools of Science • Pages 21–23• (GDCE) Complete Worksheet • Pages 31–34 • (TG)			
	Day 4	Read Special Feature: Lord Kelvin • Pages 24–25 • (GDCE)			
	Day 5				
Week 2	Day 6	Read Lesson 4: The Metric System • Pages 26–29 • (GDCE) Complete Worksheet • Pages 35–36 • (TG)			
	Day 7	Complete **Properties of Matter Quiz 1** (Lessons 1–4) Pages 339–340 • (TG)			
	Day 8	Properties of Matter Unit 2: Measuring Matter Read Lesson 5: Mass vs. Weight • Pages 31–33 • (GDCE) Complete Worksheet • Pages 37–39 • (TG)			
	Day 9	Read Lesson 6: Conservation of Mass • Pages 34–36 • (GDCE) Complete Worksheet • Pages 41–42 • (TG)			
	Day 10				
Week 3	Day 11	Read Lesson 7: Volume • Pages 37–39 • (GDCE) Complete Worksheet • Pages 43–44 • (TG)			
	Day 12	Read Lesson 8: Density • Pages 40–41 • (GDCE) Complete Worksheet • Pages 45–47 • (TG)			
	Day 13	Read Lesson 9: Buoyancy • Pages 42–44 • (GDCE) Complete Worksheet • Pages 49–50 • (TG)			
	Day 14	Read Special Feature: James Clerk Maxwell • Page 45 • (GDCE)			
	Day 15				
Week 4	Day 16	Complete **Properties of Matter Quiz 2** (Lessons 5-9) Pages 341–342 • (TG)			
	Day 17	Properties of Matter Unit 3: States of Matter Read Lesson 10: Physical & Chemical Properties Pages 47–48 • (GDCE) Complete Worksheet • Pages 51–52 • (TG)			
	Day 18	Read Lesson 11: States of Matter • Pages 49–52 • (GDCE) Complete Worksheet • Pages 53–54 • (TG)			
	Day 19	Read Lesson 12: Solids • Pages 53–55 • (GDCE) Complete Worksheet • Pages 55–56 • (TG)			
	Day 20				

Date	Day	Assignment	Due Date	✓	Grade
Week 5	Day 21	Read Lesson 13: Liquids • Pages 56–58 • (GDCE) Complete Worksheet • Pages 57–58 • (TG)			
	Day 22	Read Lesson 14: Gases • Pages 59–61 • (GDCE) Complete Worksheet • Pages 59–60 • (TG)			
	Day 23	Read Lesson 15: Gas Laws • Pages 62–64 • (GDCE) Complete Worksheet • Pages 61–62 • (TG)			
	Day 24	Read Special Feature: Robert Boyle • Page 65 • (GDCE)			
	Day 25				
Week 6	Day 26	Complete **Properties of Matter Quiz 3** (Lessons 10-15) Pages 343–344 • (TG)			
	Day 27	Properties of Matter Unit 4: Classifying Matter Read Lesson 16: Elements • Pages 67–70 • (GDCE) Complete Worksheet • Pages 63–64 • (TG)			
	Day 28	Read Special Feature: William Prout • Pages 71–72 • (GDCE)			
	Day 29	Read Lesson 17: Compounds • Pages 73–75 • (GDCE) Complete Worksheet • Pages 65–66 • (TG)			
	Day 30				
Week 7	Day 31	Read Lesson 18: Water • Pages 76–78 • (GDCE) Complete Worksheet • Pages 67–69 • (TG)			
	Day 32	Read Lesson 19: Mixtures • Pages 79–81 • (GDCE) Complete Worksheet • Pages 71–72 • (TG)			
	Day 33	Read Lesson 20: Air • Pages 82–84 • (GDCE) Complete Worksheet • Pages 73–74 • (TG)			
	Day 34	Read Lesson 21: Milk & Cream • Pages 85–88 • (GDCE) Complete Worksheet • Pages 75–76 • (TG)			
	Day 35				
Week 8	Day 36	Complete **Properties of Matter Quiz 4** (Lessons 16–21) Pages 345–346 • (TG)			
	Day 37	Properties of Matter Unit 5: Solutions Read Lesson 22: Solutions • Pages 90–91 • (GDCE) Complete Worksheet • Pages 77–80 • (TG)			
	Day 38	Read Lesson 23: Suspensions • Pages 92–94 • (GDCE) Complete Worksheet • Pages 81–82 • (TG)			
	Day 39	Read Lesson 24: Solubility • Pages 95–97 • (GDCE) Complete Worksheet • Pages 83–85 • (TG)			
	Day 40				
Week 9	Day 41	Read Lesson 25: Soft Drinks • Pages 98–100 • (GDCE) Complete Worksheet • Pages 87–88 • (TG)			
	Day 42	Read Lesson 26: Concentration • Pages 101–103 • (GDCE) Complete Worksheet • Pages 89–91 • (TG)			
	Day 43	Read Lesson 27: Seawater • Pages 104–106 • (GDCE) Complete Worksheet • Pages 93–94 • (TG)			
	Day 44	Read Special Feature: Desalination of Water • Page 107 • (GDCE)			
	Day 45				

Date	Day	Assignment	Due Date	✓	Grade
		First Semester-Second Quarter			
Week 1	Day 46	Read Lesson 28: Water Treatment • Pages 108–110 • (GDCE) Complete Worksheet • Pages 95–96 • (TG)			
	Day 47	Complete **Properties of Matter Quiz 5** (Lessons 22–28) Pages 347–348 • (TG)			
	Day 48	Properties of Matter Unit 6: Food Chemistry Read Lesson 29: Food Chemistry • Pages 112–113 • (GDCE) Complete Worksheet • Pages 97–98 • (TG)			
	Day 49	Read Special Feature: Genetically Modified Foods Pages 114–115 • (GDCE)			
	Day 50				
Week 2	Day 51	Read Lesson 30: Chemical Analysis of Food Pages 116–118 • (GDCE) Complete Worksheet • Pages 99–104 • (TG)			
	Day 52	Read Lesson 31: Flavors • Pages 119–121 • (GDCE) Complete Worksheet • Pages 105–106 • (TG)			
	Day 53	Read Special Feature: Chocolate & Vanilla • Pages 122–123 (GDCE)			
	Day 54	Read Lesson 32: Additives • Pages 124–126 • (GDCE) Complete Worksheet • Pages 107–109 • (TG)			
	Day 55				
Week 3	Day 56	Read Lesson 33: Bread • Pages 127–129 • (GDCE) Complete Worksheet • Pages 111–113 • (TG)			
	Day 57	Read Special Feature: Bread through the Centuries Pages 130–131 • (GDCE)			
	Day 58	Read Lesson 34: Identification of Unknown Substances: Final Project • Pages 132–133 • (GDCE) Complete Worksheet • Pages 115–118 • (TG)			
	Day 59	Complete **Properties of Matter Quiz 6** (Lessons 29–33) Pages 349–350 • (TG)			
	Day 60	Read Lesson 35: Conclusion • Pages 134–135 • (GDCE) Complete Worksheet • Page 119 • (TG)			
Week 4	Day 61	Complete **Properties of Matter Final Exam** (Lessons 1–34) Pages 351–353 • (TG)			
	Day 62	Properties of Ecosystems Unit 1: Introduction to Ecosystems Read Lesson 1: What Is an Ecosystem? • Pages 142–144 • (GDCE) Complete Worksheet • Pages 123–126 • (TG)			
	Day 63	Read Special Feature: Garden of Eden • Pages 145–146 • (GDCE)			
	Day 64	Read Lesson 2: Niches • Pages 147–149 • (GDCE) Complete Worksheet • Pages 127–128 • (TG)			
	Day 65				

Date	Day	Assignment	Due Date	✓	Grade
Week 5	Day 66	Read Lesson 3: Food Chains • Pages 150–152 • (GDCE) Complete Worksheet • Pages 129–130 • (TG)			
	Day 67	Read Lesson 4: Scavengers & Decomposers • Pages 153–155 • (GDCE) Complete Worksheet • Pages 131–132 • (TG)			
	Day 68	Read Lesson 5: Relationships among Living Things Pages 156–158 • (GDCE) Complete Worksheet • Pages 133–135 • (TG)			
	Day 69	Read Lesson 6: Oxygen & Water Cycles • Pages 159–161 • (GDCE) Complete Worksheet • Pages 137–138 • (TG)			
	Day 70				
Week 6	Day 71	Complete **Properties of Ecosystems Quiz 1** (Lessons 1–6) Pages 357–358 • (TG)			
	Day 72	Properties of Ecosystems Unit 2: Grasslands & Forests Read Lesson 7: Biomes around the World • Pages 163–166 • (GDCE) Complete Worksheet • Pages 139–142 • (TG)			
	Day 73	Read Special Feature: Alexander von Humboldt Pages 167–168 • (GDCE)			
	Day 74	Read Lesson 8: Grasslands • Pages 169–172 • (GDCE) Complete Worksheet • Pages 143–146 • (TG)			
	Day 75				
Week 7	Day 76	Read Lesson 9: Forests • Pages 173–175 • (GDCE) Complete Worksheet • Pages 147–149 • (TG)			
	Day 77	Read Lesson 10: Temperate Forests • Pages 176–178 • (GDCE) Complete Worksheet • Pages 151–156 • (TG)			
	Day 78	Read Lesson 11: Tropical Rainforests • Pages 179–181 • (GDCE) Complete Worksheet • Pages 157–158 • (TG)			
	Day 79	Complete **Properties of Ecosystems Quiz 2** (Lessons 7–11) Pages 359–360 • (TG)			
	Day 80				
Week 8	Day 81	Properties of Ecosystems Unit 3: Aquatic Ecosystems Read Lesson 12: The Ocean • Pages 183–186 • (GDCE) Complete Worksheet • Pages 159–161 • (TG)			
	Day 82	Read Lesson 13: Coral Reefs • Pages 187–189 • (GDCE) Complete Worksheet • Pages 163–164 • (TG)			
	Day 83	Read Lesson 14: Beaches • Pages 190–192 • (GDCE) Complete Worksheet • Pages 165–167 • (TG)			
	Day 84	Read Lesson 15: Estuaries • Pages 193–195 • (GDCE) Complete Worksheet • Pages 169–171 • (TG)			
	Day 85				
Week 9	Day 86	Read Lesson 16: Lakes & Ponds • Pages 196–198 • (GDCE) Complete Worksheet • Pages 173–177 • (TG)			
	Day 87	Read Lesson 17: Rivers & Streams • Pages 199–200 • (GDCE) Complete Worksheet • Pages 179–183 • (TG)			
	Day 88	Read Special Feature: The Amazon River • Pages 201–202 • (GDCE)			
	Day 89	Complete **Properties of Ecosystems Quiz 3** (Lessons 12–17) Pages 361–362 • (TG)			
	Day 90				
		Mid-Term Grade			

Second Semester Suggested Daily Schedule

Date	Day	Assignment	Due Date	✓	Grade
		Second Semester-Third Quarter			
Week 1	Day 91	Properties of Ecosystems Unit 4: Extreme Ecosystems Read Lesson 18: Tundra • Pages 204–207 • (GDCE) Complete Worksheet • Pages 185–187 • (TG)			
	Day 92	Read Special Feature: Robert Peary Pages 208–209 • (GDCE)			
	Day 93	Read Lesson 19: Deserts • Pages 210–213 • (GDCE) Complete Worksheet • Pages 189–191 • (TG)			
	Day 94	Read Lesson 20: Oases • Pages 214–216 • (GDCE) Complete Worksheet • Pages 193–194 • (TG)			
	Day 95				
Week 2	Day 96	Read Lesson 21: Mountains • Pages 217–220 • (GDCE) Complete Worksheet • Pages 195–197 • (TG)			
	Day 97	Read Lesson 22: Chaparral • Pages 221–223 • (GDCE) Complete Worksheet • Pages 199–201 • (TG)			
	Day 98	Read Lesson 23: Caves • Pages 224–227 • (GDCE) Complete Worksheet • Pages 203–205 • (TG)			
	Day 99	Complete **Properties of Ecosystems Quiz 4** (Lessons 18–23) Pages 363–364 • (TG)			
	Day 100				
Week 3	Day 101	Properties of Ecosystems Unit 5: Animal Behaviors Read Lesson 24: Seasonal Behaviors • Pages 229–232 • (GDCE) Complete Worksheet • Pages 207–208 • (TG)			
	Day 102	Read Lesson 25: Animal Defenses • Pages 233–235 • (GDCE) Complete Worksheet • Pages 209–210 • (TG)			
	Day 103	Read Lesson 26: Adaptation • Pages 236–238 • (GDCE) Complete Worksheet • Pages 211–212 • (TG)			
	Day 104	Read Lesson 27: Balance of Nature • Pages 239–242 • (GDCE) Complete Worksheet • Pages 213–215 • (TG)			
	Day 105				
Week 4	Day 106	Read Special Feature: Eugene P. Odum • Page 243 • (GDCE)			
	Day 107	Complete **Properties of Ecosystems Quiz 5** (Lessons 24–27) Pages 365–366 • (TG)			
	Day 108	Properties of Ecosystems Unit 6: Ecology & Conservation Read Lesson 28: Man's Impact on the Environment Pages 245–247 • (GDCE) Complete Worksheet • Pages 217–218 • (TG)			
	Day 109	Read Lesson 29: Endangered Species • Pages 248–251 • (GDCE) Complete Worksheet • Pages 219–220 • (TG)			
	Day 110				

Date	Day	Assignment	Due Date	✓	Grade
Week 5	Day 111	Read Special Feature: Theodore Roosevelt Pages 252–253 • (GDCE)			
	Day 112	Read Lesson 30: Pollution • Pages 254–257 • (GDCE) Complete Worksheet • Pages 221–223 • (TG)			
	Day 113	Read Lesson 31: Acid Rain • Pages 258–260 • (GDCE) Complete Worksheet • Pages 225–226 • (TG)			
	Day 114	Read Lesson 32: Global Warming • Pages 261–264 • (GDCE) Complete Worksheet • Pages 227–230 • (TG)			
	Day 115				
Week 6	Day 116	Read Lesson 33: What Can You Do? • Pages 265–267 • (GDCE) Complete Worksheet • Pages 231–233 • (TG)			
	Day 117	Read Lesson 34: Reviewing Ecosystems: Final Project Page 268 • (GDCE) Complete Worksheet • Page 235 • (TG)			
	Day 118	Complete **Properties of Ecosystems Quiz 6** (Lessons 28–33) Pages 367–368 • (TG)			
	Day 119	Read Lesson 35: Conclusion • Page 269 • (GDCE) Complete Worksheet • Page 237 • (TG)			
	Day 120				
Week 7	Day 121	Complete **Properties of Ecosystems Final Exam** (Lessons 1–34) Pages 369–373 • (TG)			
	Day 122	Properties of Atoms & Molecules Unit 1: Atoms & Molecules Read Lesson 1: Introduction to Chemistry Pages 278–279 • (GDCE) Complete Worksheet • Pages 241–242 • (TG)			
	Day 123	Read Lesson 2: Atoms • Pages 280–282 • (GDCE) Complete Worksheet • Pages 243–246 • (TG)			
	Day 124	Read Lesson 3: Atomic Mass • Pages 283–284 • (GDCE) Complete Worksheet • Pages 247–250 • (TG)			
	Day 125				
Week 8	Day 126	Read Special Feature: Madame Curie • Pages 285–286 • (GDCE)			
	Day 127	Read Lesson 4: Molecules • Pages 287–289 • (GDCE) Complete Worksheet • Pages 251–254 • (TG)			
	Day 128	Complete **Properties of Atoms & Molecules Quiz 1** (Lessons 1–4) • Pages 377–378 • (TG)			
	Day 129	Properties of Atoms & Molecules Unit 2: Elements Read Lesson 5: Periodic Table of the Elements Pages 291–294 • (GDCE) Complete Worksheet • Pages 255–256 • (TG)			
	Day 130				
Week 9	Day 131	Read Special Feature: Development of the Periodic Table Page 295 (GDCE)			
	Day 132	Read Lesson 6: Metals • Pages 296–298 • (GDCE) Complete Worksheet • Pages 257–258 • (TG)			
	Day 133	Read Lesson 7: Nonmetals • Pages 299–301 • (GDCE) Complete Worksheet • Pages 259–260 • (TG)			
	Day 134	Read Lesson 8: Hydrogen • Pages 302–304 • (GDCE) Complete Worksheet • Pages 261–262 • (TG)			
	Day 135				

Date	Day	Assignment	Due Date	✓	Grade
		Second Semester-Fourth Quarter			
Week 1	Day 136	Read Lesson 9: Carbon • Pages 305–307 • (GDCE) Complete Worksheet • Pages 263–264 • (TG)			
	Day 137	Read Lesson 10: Oxygen • Pages 308–310 • (GDCE) Complete Worksheet • Pages 265–266 • (TG)			
	Day 138	Complete **Properties of Atoms & Molecules Quiz 2** (Lessons 5-10) • Pages 379–380 • (TG)			
	Day 139	Properties of Atoms & Molecules Unit 3: Bonding Read Lesson 11: Ionic Bonding • Pages 312–315 • (GDCE) Complete Worksheet • Pages 267–269 • (TG)			
	Day 140				
Week 2	Day 141	Read Lesson 12: Covalent Bonding • Pages 316–318 • (GDCE) Complete Worksheet • Pages 271–273 • (TG)			
	Day 142	Read Lesson 13: Metallic Bonding • Pages 319–320 • (GDCE) Complete Worksheet • Pages 275–276 • (TG)			
	Day 143	Read Lesson 14: Mining & Metal Alloys Pages 321–323 • (GDCE) Complete Worksheet • Pages 277–278 • (TG)			
	Day 144	Read Special Feature: Charles Martin Hall Pages 324–325 • (GDCE)			
	Day 145				
Week 3	Day 146	Read Lesson 15: Crystals • Pages 326–329 • (GDCE) Complete Worksheet • Pages 279–280 • (TG)			
	Day 147	Read Lesson 16: Ceramics • Pages 330–332 • (GDCE) Complete Worksheet • Pages 281–282 • (TG)			
	Day 148	Complete **Properties of Atoms & Molecules Quiz 3** (Lessons 11–16) • Pages 381–382 • (TG)			
	Day 149	Properties of Atoms & Molecules Unit 4: Chemical Reactions Read Lesson 17: Chemical Reactions • Pages 334–337 • (GDCE) Complete Worksheet • Pages 283–285 • (TG)			
	Day 150				
Week 4	Day 151	Read Lesson 18: Chemical Equations • Pages 338–340 • (GDCE) Complete Worksheet • Pages 287–289 • (TG)			
	Day 152	Read Lesson 19: Catalysts • Pages 341–343 • (GDCE) Complete Worksheet • Pages 291–292 • (TG)			
	Day 153	Read Lesson 20: Endothermic & Exothermic Reactions Pages 344–346 • (GDCE) Complete Worksheet • Pages 293–295 • (TG)			
	Day 154	Complete **Properties of Atoms & Molecules Quiz 4** (Lessons 17–20) • Pages 383–384 • (TG)			
	Day 155	Properties of Atoms & Molecules Unit 5: Acids & Bases Read Lesson 21: Chemical Analysis • Pages 348–350 • (GDCE) Complete Worksheet • Pages 297–298 • (TG)			

Date	Day	Assignment	Due Date	✓	Grade
Week 5	Day 156	Read Lesson 22: Acids • Pages 351–353 • (GDCE) Complete Worksheet • Pages 299–300 • (TG)			
	Day 157	Read Lesson 23: Bases • Pages 354–356 • (GDCE) Complete Worksheet • Pages 301–302 • (TG)			
	Day 158	Read Lesson 24: Salts • Pages 357–359 • (GDCE) Complete Worksheet • Pages 303–305 • (TG)			
	Day 159	Read Special Feature: Batteries • Pages 360–361 • (GDCE)			
	Day 160				
Week 6	Day 161	Complete **Properties of Atoms & Molecules Quiz 5** (Lessons 21–24) • Pages 385–386 • (TG)			
	Day 162	Properties of Atoms & Molecules Unit 6: Biochemistry Read Lesson 25: Biochemistry • Pages 363–366 • (GDCE) Complete Worksheet • Pages 307–309 • (TG)			
	Day 163	Read Lesson 26: Decomposers • Pages 367–369 • (GDCE) Complete Worksheet • Pages 311–313 • (TG)			
	Day 164	Read Lesson 27: Chemicals in Farming • Pages 370–372 • (GDCE) Complete Worksheet • Pages 315–316 • (TG)			
	Day 165				
Week 7	Day 166	Read Lesson 28: Medicines • Pages 373–375 • (GDCE) Complete Worksheet • Pages 317–318 • (TG) • Read Special Feature: Alexander Fleming • Pages 376–377 • (GDCE)			
	Day 167	Complete **Properties of Atoms & Molecules Quiz 6** (Lessons 25–28) • Pages 387–388 • (TG)			
	Day 168	Properties of Atoms & Molecules Unit 7: Applications of Chemistry • Read Lesson 29: Perfumes • Pages 379–381 • (GDCE) Complete Worksheet • Pages 319–320 • (TG)			
	Day 169	Read Lesson 30: Rubber • Pages 382–385 • (GDCE) Complete Worksheet • Pages 321–322 • (TG)			
	Day 170				
Week 8	Day 171	Read Special Feature: Charles Goodyear Pages 386–387 • (GDCE)			
	Day 172	Read Lesson 31: Plastics • Pages 388–390 • (GDCE) Complete Worksheet • Pages 323–325 • (TG)			
	Day 173	Read Lesson 32: Fireworks • Pages 391–393 • (GDCE) Complete Worksheet • Pages 327–328 • (TG)			
	Day 174	Read Lesson 33: Rocket Fuel • Pages 394–396 • (GDCE) Complete Worksheet • Pages 329–330 • (TG)			
	Day 175				
Week 9	Day 176	Read Lesson 34: Fun with Chemistry: Final Project Pages 397–399 • (GDCE) Complete Worksheet • Pages 331–333 • (TG)			
	Day 177	Complete **Properties of Atoms & Molecules Quiz 7** (Lessons 29–33) • Pages 389–390 • (TG)			
	Day 178	Read Lesson 35: Conclusion • Page 400 • (GDCE) Complete Worksheet • Page 335 • (TG)			
	Day 179	Complete **Properties of Atoms & Molecules Final Exam** (Lessons 1–34) • Pages 391–393 • (TG)			
	Day 180				
		Final Grade			

Matter Worksheets

for Use with

Properties of Matter

(*God's Design: Chemistry & Ecology*)

| God's Design: Chemistry & Ecology | Properties of Matter | Day 1 | Unit 1 Lesson 1 | Name |

1 Introduction to Experimental Science

Learning about matter

🧪 Supply list – Chemistry is fun

☐ Wooden spoon ☐ Pencil ☐ Butter
☐ Metal spoon ☐ Butter knife ☐ Stopwatch
☐ Plastic ruler ☐ Large cup of hot water ☐ Copy of "Conducting Heat Experiment" Worksheet

🧪 Chemistry is fun

1. Did the butter melt fastest on the item you expected to conduct the heat the best?

2. Which items actually conducted heat the best? Which ones conducted heat the slowest?

🎖 Operational science vs. origins science

Note: Challenge topics can be discussed or answered orally. Or, the student can write his/her answer.
Explain to your teacher the difference between operational science vs. origins science.

🧠 What did we learn?

1. What is matter?

2. What do chemists study?

3. What is an experiment?

🚀 Taking it further

1. Why is it important to study chemistry?

2. What are two things you need to know before conducting an experiment?

Lesson 1 **Properties of Matter** // 25

Name _____ Date_____

🧪 Conducting Heat Experiment Worksheet

Which items listed on the chart below do you think will conduct heat the fastest? Write your hypothesis below and then perform the experiment in your student manual.

Hypothesis

I think the order for the conductivity of the items will be:

_____ (Fastest heat conductor)

_____

_____

_____

_____ (Slowest heat conductor)

Observations

Item	Time to melt butter
Metal spoon	
Wooden spoon	
Plastic ruler	
Pencil	
Butter knife	

Conclusions

The actual order for the conductivity of the items was:

_____ (Fastest heat conductor)

_____

_____

_____

_____ (Slowest heat conductor)

| God's Design: Chemistry & Ecology | Properties of Matter | Day 2 | Unit 1 Lesson 2 | Name |

2 The Scientific Method

How do scientists do it?

🧪 Supply list – Using the scientific method

- ☐ 3 empty plastic bottles
- ☐ Masking tape
- ☐ Yeast
- ☐ Marker
- ☐ Sugar
- ☐ Molasses
- ☐ 3 identical balloons
- ☐ Thermometer
- ☐ Measuring cup and spoons
- ☐ Warm water
- ☐ Cloth tape measure or string
- ☐ Copy of "Scientific Method" Worksheet

🏅 Design your own experiment

Describe the experiment you designed either on your own or with one of the suggested topics. Show your teacher the data sheet where you recorded data from this experiment, how you are controlling your variables, and the conclusion for your experiment.

🧠 What did we learn?

1. What is the overall job of a scientist?

2. What are some areas that cannot be studied by science?

3. What are the five steps of the scientific method?

Taking it further

1. Why was it necessary to have bottle number 1 in the experiment?

2. What other sweeteners could you try in your experiment?

3. What sweeteners were used in the bread at your house?

4. Why do you think that sweetener was used?

Name _____ Date _____

🧪 Scientific Method Worksheet

Write your hypothesis below and then perform the experiment in your student manual.

Hypothesis

Do you think sugar or molasses will produce the most gas? _____

Circumference of Balloon

Time	Bottle 1 (no sweetener)	Bottle 2 (sugar)	Bottle 3 (molasses)

Which balloon had the most gas after 1 hour? _____

Did the bottle with your chosen sweetener produce the most gas? _____

Did this support your hypothesis? _____

Which sweetener would you use to make your bread? _____

Why might someone choose to use a sweetener that does not produce the most gas? _____
_____

Lesson 2 **Properties of Matter** // 29

| God's Design: Chemistry & Ecology | Properties of Matter | Day 3 | Unit 1 Lesson 3 | Name |

3 Tools of Science

Using the right tool for the job

🧪 Supply list – Learning to use your tools

☐ Thermometer
☐ Masking tape
☐ Liquid measuring cup
☐ Marker
☐ Small box
☐ Tennis ball
☐ 2 cups
☐ Metric ruler or meter stick
☐ Digital stop watch
☐ Copy of "Scientific Tools" Worksheet

🏅 Supplies for Challenge (if available) – Magnifying objects

☐ Microscope
☐ Prepared slides
☐ Telescope

🧠 What did we learn?

1. What is the main thing a scientist does as he/she studies the physical world?

2. What are the two types of observations that a scientist can make?

3. What is the main problem with qualitative observations?

4. What are some scientific tools used for quantitative observations?

🚀 Taking it further

1. What qualitative observations might you make when observing the experiment in lesson 1?

2. What quantitative observations might you make when observing the experiment in lesson 1?

Name _____   Date _____

🧪 Scientific Tools Worksheet

Test 1

Use masking tape and a marker to label two cups as Cup 1 and Cup 2. Fill Cup 1 with hot tap water and Cup 2 with cold tap water.

Qualitative observations

Use your five senses to describe the water in each cup.

Cup 1: _____

Cup 2: _____

Quantitative measurement

Use a thermometer to measure the temperature of the water in each cup.

The temperature is:

Cup 1:   Cup 2:

Test 2

Determine which cup has more water in it.

Qualitative observations

Which cup appears to have more water in it? _____

Quantitative measurement

Pour the contents of Cup 1 into a liquid measuring cup and record the amount of water below. Empty the measuring cup and repeat for Cup 2.

Amount of water in Cup 1:

Amount of water in Cup 2:

Which cup had the most water in it? _____

How much more water did it have? _____

Lesson 3 **Properties of Matter** 🔗 33

Test 3

Observe a small box.

Qualitative observations

Describe the size, shape, texture, and color of the box.

_____

Quantitative measurement

Use a ruler or meter stick to measure the box.

Length: Height: Width:

Test 4

Hold a tennis ball at waist height and drop it. Next, hold the tennis ball as high as you can reach and drop it. From which height did the ball reach the ground the fastest?

Qualitative observations

Describe how long it took the ball to fall each time.

From waist high: From high up:

Quantitative measurement

Repeat the experiment using a digital stop watch to measure the length of time it takes for the ball to reach the ground after it is released.

From waist high: From high up:

Conclusions

Which type of measurements gave you a more accurate answer? _____

Which type of measurements do you think are more useful? _____

Is it always necessary to make quantitative measurements? _____

| God's Design: Chemistry & Ecology | Properties of Matter | Day 6 | Unit 1 Lesson 4 | Name |

4 The Metric System

Standard units

🧪 Supply list – Using metric units

☐ Metric measuring cup
☐ Meter stick
☐ 1-liter container
☐ Pencil
☐ Paper clip

🧪 Using metric units

Note: You can choose to do one or more of the experiments. You can discuss your results for Activities 1–3. If the Activity 4 Procedure is being done, you can have the student answer the questions verbally or write the answers below. The conversion chart is on page 27 of the student book.

1. If you pour 1,000 ml of water into a bottle, how many liters of water do you have?

2. If you weigh 20 kg, how many grams do you weigh?

3. If it is 40 hectometers from your house to your best friend's house, how many meters must you walk to get from your house to his/hers?

4. If your pet hamster is 60 mm long, how long is it in cm?

5. If you have 5 dekagrams of chocolate to share, how many decigrams do you have?

🎖 Measuring scales

Note: You can modify this challenge based on the topics in the student book for the interests of your student.

1. What are two ways that earthquakes can be measured that you discovered from your research?

What did we learn?

1. What are some units used to measure length in the Old English/American measuring system?

2. What is the unit used to measure length in the metric system?

3. What metric unit is used for measuring mass?

4. What metric unit is used for measuring liquid volume?

5. Why do scientists use the metric system instead of another measuring system?

Taking it further

1. What metric unit would be best to use to measure the distance across a room?

2. What metric unit would you use to measure the distance from one town to another?

3. What metric unit would you use to measure the width of a hair?

| God's Design: Chemistry & Ecology | Properties of Matter | Day 8 | Unit 2 Lesson 5 | Name |

5 Mass vs. Weight

What's the difference?

🧪 Supply list – Measuring mass; Measuring weight

- ☐ Ruler with holes punched in it to fit a 3-ring binder
- ☐ Rubber band or spring
- ☐ String
- ☐ 2 pencils
- ☐ 3 paper cups
- ☐ 25 pennies
- ☐ Tape
- ☐ Hole punch
- ☐ Paper clips

Note: Be sure to save the balance for Lessons 6 and 8.

🎖 Supplies for Challenge – Mass & weight units

- ☐ Copy of "Mass and Weight Units" Worksheet

🧠 What did we learn?

1. What is the difference between mass and weight?

2. How do you measure mass?

3. How do you measure weight?

🚀 Taking it further

1. What would your weight be in outer space?

2. What would your mass be in outer space?

3. Name a place in the universe where you might go to increase your weight without changing your mass.

Name _____   Date_____

🏅 Mass & Weight Units Worksheet

For each sentence below, mark whether the sentence is referring to the mass or the weight of the object discussed.

1. _____ The ball is 15 newtons.

2. _____ The girl had a 200-gram book.

3. _____ The 2-slug man had a green shirt.

4. _____ Place the box on the spring scale.

5. _____ He lifted the 10-pound dog.

6. _____ You can use the pan balance to measure the chemical.

7. _____ Johnny is 20 pounds heavier than Sally.

8. _____ Can you carry 10 kilograms?

9. _____ I have 1 slug of nails.

10. _____ Place the gram pieces on the right side of the balance.

11. _____ I ate 30 grams of grapes at the Smith's house.

12. _____ How many newtons can that bookshelf hold?

Lesson 5 **Properties of Matter**

6 Conservation of Mass

Where does it go?

Supply list – Changing form without losing mass

☐ 2 paper cups
☐ Balance from lesson 5
☐ 3 or 4 sugar cubes
☐ Spoon

Supplies for Challenge – Mass of gases

☐ Bottle with a narrow neck
☐ Gram scale
☐ Vinegar
☐ Baking soda
☐ Paper
☐ Balloon
☐ Copy of "Conservation of Mass" Worksheet

What did we learn?

1. What is the law of conservation of mass?

2. How is the mass of water changed when it turns to ice?

Taking it further

1. If you start with 10 grams of water and you boil it until there is no water left in the pan, what happened to the water?

2. Why is the law of conservation of mass important to understanding the beginning of the world?

Name _____   Date _____

Conservation of Mass Worksheet

Perform the experiment in your student manual and record your observations below.

	Mass before reaction	Mass after reaction	Change in mass
Open bottle, paper, baking soda, vinegar			
Bottle with balloon, baking soda, vinegar			

1. How much did the mass change when the bottle was open? _____

2. Why is the mass less after the first reaction? _____
   _____

3. Where did the missing mass go? _____

4. How much did the mass change after the reaction when the bottle was covered? _____
   _____

5. If the mass is different in the second reaction, give some possible explanations.
   _____
   _____

Even though many gases are not visible, this does not mean that they have no mass. Carbon dioxide, oxygen, nitrogen, and many other gases have mass but are not visible to the human eye.

| God's Design: Chemistry & Ecology | Properties of Matter | Day 11 | Unit 2 Lesson 7 | Name |

7 Volume

How much space does it take up?

🧪 Supply list – Measuring volume

☐ Meter stick
☐ Small box
☐ Metric ruler
☐ Drinking glass
☐ Liquid measuring cup
☐ Irregularly shaped object

🏅 Supplies for Challenge – Calculating volume

☐ Box of dry food (crackers, cereal, etc.)
☐ Tennis ball
☐ Soup or canned food
☐ Ice cream cone
☐ Die (6-sided)
☐ Copy of "Calculating Volume" Worksheet

🧠 What did we learn?

1. What is volume?

2. Does air have volume?

🚀 Taking it further

1. If you have a cube that is 10 centimeters on each side, what would its volume be?

2. Why is volume important to a scientist?

Lesson 7 **Properties of Matter** // 43

Name _____   Date_____

🎖 Calculating Volume Worksheet

Volume of a cube = Side x Side x Side
Volume of a rectangle = Length x Width x Height
Volume of a sphere = (4/3) π Radius x Radius x Radius   Where π = approximately 3.14
Volume of a cylinder = π Radius x Radius x Height
Volume of a cone = (1/3) π Radius x Radius x Height

Use the formulas above to calculate the volume of each of the following items.

(Note: You can use anything that is shaped like the items below if you do not have the exact items; for example, you can make a cone out of paper if you do not have an ice cream cone available.)

Box of crackers or other dry food:

Tennis ball:

(Note: It can be difficult to accurately measure the radius of a ball. It is much easier to measure the circumference with a piece of string and then calculate the radius using C = 2 x π x Radius, so the radius is equal to R= C/2π)

Can of soup or other canned food:

Ice cream cone (without the ice cream):

Die (6-sided for playing games):

8 Density

Does it feel heavy?

🧪 Supply list – Measuring density

☐ Ping-pong ball
☐ Golf ball
☐ Balance from lesson 5
☐ Pennies
☐ Paper clips
☐ Liquid measuring cup

🧪 Measuring density

1. How did the mass of the golf ball compare to the mass of the ping-pong ball?

2. How did the volume of the golf ball compare to the volume of the ping-pong ball?

3. Which ball has a higher density?

🏅 Supplies for Challenge – Density experiment

☐ Metal spoon
☐ Marble
☐ Eraser
☐ Plastic cap
☐ Quarter
☐ Copy of "Density Experiment" Worksheet

🧠 What did we learn?

1. What is the definition of density?

Lesson 8 Properties of Matter 45

2. If two substances with the same volume have different densities, how can you tell which one is the densest?

🚀 Taking it further

1. If you have two unknown substances that both appear to be silvery colored, how can you tell if they are the same material?

2. If two objects have the same density and the same size, what will be true about their masses?

3. If you suspect that someone is trying to pass off a gold-plated bar of lead as a solid gold bar, how can you test your theory?

4. Why does the ping-pong ball have a lower density than the golf ball?

Name _____   Date_____

ⓠ Density Experiment Worksheet

Look at the list of items in the chart below. Based on what you know about each item, make a hypothesis about their relative densities by listing them on the lines below from least dense to most dense.

Hypothesis

I think that the relative densities are:

1. _____ (least dense)

2. _____

3. _____

4. _____

5. _____ (most dense)

Observations

Now measure the mass and volume of each item and record them on the chart below. Finally, calculate the density of each item and record it on the chart as well.

Item	Mass	Volume	Density
Metal spoon			
Marble			
Eraser			
Plastic cap			
Quarter			

Conclusions

The actual order of relative density from least to most dense for the items above is:

1. _____ (least dense)

2. _____

3. _____

4. _____

5. _____ (most dense)

Lesson 8 Properties of Matter ⁄⁄ 47

| God's Design: Chemistry & Ecology | Properties of Matter | Day 13 | Unit 2 Lesson 9 | Name |

9 Buoyancy

It floats.

🧪 Supply list – Testing buoyancy

☐ Rubbing alcohol

☐ Vegetable oil

☐ Modeling clay

☐ 2 cups

☐ Popcorn (including some unpopped kernels)

🎖 Supplies for Challenge – Buoyancy questions

☐ Plastic tub

☐ Sink

☐ Tape

☐ Water

☐ Helium balloon (Optional challenge)

☐ Metal objects (wrenches, etc.) that fit in the tub

🎖 Buoyancy questions

What were the results of your challenge?

🧠 What did we learn?

1. What is buoyancy?

2. If something is buoyant, what does that tell you about its density compared to that of the substance in which it floats?

3. Are you buoyant in water?

🚀 Taking it further

1. What are some substances that are buoyant in water besides you?

2. Based on what you observed, which is denser, water or alcohol?

3. Why is a foam swimming tube or a foam life ring able to keep a person afloat in the water?

4. Why is it important to life that ice is less dense than water?

| God's Design: Chemistry & Ecology | Properties of Matter | Day 17 | Unit 3 Lesson 10 | Name |

10 Physical & Chemical Properties

Is it something new?

🧪 Supply list – Physical & chemical changes

☐ Sauce pan ☐ Lemon juice ☐ Access to a stove
☐ Ice ☐ Baking soda
☐ Salt ☐ Cup

🧪 Physical & chemical changes

Briefly explain what you learned from your experiment. Be sure to share your observations on the physical changes and chemical changes.

🎖 Supplies for Challenge – Physical or chemical?

☐ Copy of "Physical or Chemical Properties" Worksheet

🧠 What did we learn?

1. What are some physical properties of matter?

2. What is a chemical change?

3. Give an example of a chemical change.

🚀 Taking it further

1. How can you determine if a change in matter is a physical change or a chemical change?

2. Diamond and quartz appear to have very similar physical properties. They are both clear crystalline substances. However, diamond is much harder than quartz. How would this affect each one's effectiveness as tips for drill bits?

Name _____   Date_____

🏅 Physical or Chemical Properties Worksheet

For each change or attribute listed below, write *P* if it describes a physical change or physical property, and *C* if it describes a chemical change or chemical property. To determine if it is a chemical change, ask yourself if a new substance is being formed.

1. _____ Liquid water becoming steam

2. _____ Flavor/taste

3. _____ Burning of wood/fire

4. _____ Filling a balloon with air

5. _____ Softness

6. _____ Making ice cream

7. _____ Digesting food

8. _____ Straightening a paper clip

9. _____ Cloud formation

10. _____ Rust on a piece of iron

11. _____ Separation of water into hydrogen and oxygen gases

12. _____ Dissolving sugar in water

13. _____ Photosynthesis

14. _____ Bacteria decaying dead plant matter

15. _____ Shine/luster

16. _____ A cake rising in the oven

17. _____ Cutting a piece of wood

18. _____ Bread rising

19. _____ Hardness

20. _____ Making perfume

11 States of Matter

Phase changes

Supply list – Observing phase changes

☐ Ice
☐ Small sauce pan
☐ Hand mirror
☐ Ice tray
☐ Access to stove and freezer

Observing phase changes

Explain to your teacher the results of your experiment.

Supplies for Challenge – Water density

☐ Glass jar
☐ Marker
☐ Water
☐ Access to freezer

Water density

1. What did you learn from this challenge?

What did we learn?

1. What are the three physical states of most matter?

    a.

    b.

    c.

2. What is the name for each phase change?

3. What is required to bring about a phase change in a substance?

🚀 Taking it further

1. Name several substances that are solid at room temperature.

2. Name several substances that are liquid at room temperature.

3. Name several substances that are gas at room temperature.

| God's Design: Chemistry & Ecology | Properties of Matter | Day 19 | Unit 3 Lesson 12 | Name |

12 Solids

Hard as rock

Supply list – Testing for solids

☐ Wooden block
☐ Honey
☐ Rock
☐ Metal spoon
☐ Silly Putty®

Testing for solids

What surprised you most in this experiment?

Glass: solid or liquid?

1. What did you determine from your research — is glass a solid or liquid?

What did we learn?

1. What are three characteristics of solids?

    a.

    b.

    c.

2. How do large crystals form in solids?

3. What state is the most common for the basic elements?

 Taking it further

1. Is gelatin a solid or a liquid?

| God's Design: Chemistry & Ecology | Properties of Matter | Day 21 | Unit 3 Lesson 13 | Name |

13 Liquids

Can you pour it?

🧪 Supply list – Observing viscosity

☐ Water
☐ Hand lotion
☐ Vegetable Oil
☐ Dish soap
☐ Honey
☐ Baking sheet
☐ Copy of "Viscosity" Worksheet

🎖 Supplies for Challenge – Evaporation

☐ Cup of water

🎖 Evaporation

1. Which finger becomes dry faster?

2. How does the finger you blow on feel compared to the finger you did not blow on?

🧠 What did we learn?

1. Which has more kinetic energy, a solid or a liquid?

2. What shape does a liquid have?

3. What is viscosity?

🚀 Taking it further

1. How is a liquid similar to a solid?

2. How is a liquid different from a solid?

3. How would you change a solid into a liquid?

Lesson 13 **Properties of Matter** ✐✐ 57

Name _____ Date_____

🧪 Viscosity Worksheet

To understand that some liquids can be thicker than others, we need to study several different liquids. For each of the liquids on the chart below, observe the liquid in its original container. Then feel a small amount of each liquid with your fingers and record your observations below.

Liquid	How it looks in the container	How it feels when I touch it
Water		
Vegetable oil		
Dish soap		
Hand lotion		
Honey		

Based on your observations, list the five liquids in the order you think would be from thinnest to thickest.

1. _____ (Thinnest—low viscosity)

2. _____

3. _____

4. _____

5. _____ (Thickest—high viscosity)

Now test your hypothesis by placing ½ teaspoon of each liquid in drops evenly spaced across the short edge of a baking sheet. Lift the edge of the baking sheet several inches and observe as the liquids flow down the sheet.

List the liquids in order of how fast they reached the bottom of the baking sheet. Some liquids may not reach the bottom. List them according to how far they flowed down the sheet.

1. _____ (Fastest/thinnest—lowest viscosity)

2. _____

3. _____

4. _____

5. _____ (Slowest/thickest—highest viscosity)

Was your hypothesis correct? (Did you list them in the correct order?) _____

Try to find a liquid that is thicker than the thickest liquid you just tested. The thickest liquid I could find was

_____.

58 // God's Design: Chemistry & Ecology

14 Gases

Lighter than air?

🧪 Supply list – Observing air pressure

☐ 2 tennis balls (Note: place 1 of the balls in the freezer about 30 minutes before you plan to use it.)

☐ Access to a freezer

🧪 Observing air pressure

1. Which ball bounced highest? Why?

🎖 Supplies for Challenge – Diffusion

☐ Perfume

☐ Cup

☐ Step ladder

🎖 Diffusion

Discuss the process of diffusion.

🧠 What did we learn?

1. When is a substance called a gas?

2. What is the shape of a gas?

3. In which state of matter are the molecules moving the fastest?

4. What is atmospheric pressure?

🚀 Taking it further

1. How is a gas similar to a liquid?

2. How is a gas different from a liquid?

3. Why is it necessary that a spacesuit be pressurized in outer space?

| God's Design: Chemistry & Ecology | Properties of Matter | Day 23 | Unit 3 Lesson 15 | Name |

15 Gas Laws

Rules to live by

🧪 Supply list – Hot and cold gas

☐ Empty plastic 1-gallon milk carton
☐ Balloon
☐ Microwave oven
☐ Oven mitts
☐ Cloth tape measure (or string and a ruler)
☐ Access to a freezer

🎖 Supplies for Challenge – Charles's law

☐ Small plastic bottle
☐ Dish soap
☐ Hot and cold water
☐ 2 small bowls

🎖 Charles's law

1. What happened when you placed the bottle in the hot water?

2. What happened when you moved the bottle to the cold water? Why did this occur?

🧠 What did we learn?

1. If temperature remains constant, what happens to the volume of a gas when the pressure is increased?

2. If pressure remains constant, what happens to the volume of a gas when the temperature is increased?

3. What are two different ways to increase the volume of a gas?
   a.

   b.

🚀 Taking it further

1. Why might you need to check the air in your bike tires before you go for a ride on a cold day?

2. Why do you think increasing pressure decreases the volume of a gas?

3. Why do you think increasing temperature increases the volume of a gas?

4. What might happen to the volume of a gas when the pressure is increased and the temperature is increased at the same time?

| God's Design: Chemistry & Ecology | Properties of Matter | Day 27 | Unit 4 Lesson 16 | Name |

16 Elements

The basic building blocks

Supply list – Classification exercise
☐ Jigsaw puzzle

Supplies for Challenge – Periodic table
☐ Copy of "Learning about Elements" Worksheet

What did we learn?

1. What is an element?

2. What is a compound?

3. What is a mixture?

Taking it further

1. If a new element were discovered and named *newmaterialium*, would you expect it to be a metal or a nonmetal?

2. Is salt an element, a compound, or a mixture?

3. Is a soft drink an element, a compound, or a mixture?

Lesson 16 **Properties of Matter** // 63

Name _____   Date_____

🎖 Learning about Elements Worksheet

Use the periodic table of the elements (student manual, lesson 16) to answer the following questions. What are the symbol and atomic number for each of the following elements?

Name	Symbol	Atomic number
Hydrogen		
Oxygen		
Aluminum		
Silicon		
Mercury		

Put the following elements in the correct column on the chart below.

Sodium	Phosphorus	Barium	Antimony
Nitrogen	Fluorine	Potassium	Boron
Germanium	Neon	Calcium	Chlorine
Gold	Polonium	Arsenic	Silver

Metal	Metalloid	Nonmetal

Place the following elements in order from smallest atomic number to largest atomic number.

Iron	Argon	Bismuth	Sulfur
Copper	Magnesium	Platinum	Radon

_____ (smallest atomic number)
_____
_____
_____
_____
_____
_____
_____ (largest atomic number)

64 // God's Design: Chemistry & Ecology

God's Design: Chemistry & Ecology | Properties of Matter | Day 29 | Unit 4 Lesson 17 | Name

17 Compounds

Making new substances

Supply list – Electrolysis of water

☐ 2 small jars (small baby food jars or test tubes)
☐ Copper wire (at least 3 feet)
☐ 6-volt battery (big square battery)
☐ Baking soda
☐ Sink

Electrolysis of water

1. What do you think is in each jar?

2. Which jar do you think has the hydrogen in it?

3. Why do you think the battery is needed to separate the atoms?

What did we learn?

1. What is a compound?

2. What is another name for an element?

3. What is another name for a compound?

4. Do compounds behave the same way as the atoms that they are made from?

🚀 Taking it further

1. The symbol for carbon dioxide is CO_2. What atoms combine to form this molecule?

2. The air consists of nitrogen and oxygen molecules. Is air a compound? Why or why not?

| God's Design: Chemistry & Ecology | Properties of Matter | Day 31 | Unit 4 Lesson 18 | Name |

18 Water

God's compound for life

🧪 Supply list – Hunting for water

☐ Copy of "Water, Water Everywhere" Worksheet

🏅 Supplies for Challenge – Water recycling

☐ Three 2-liter plastic soft drink bottles
☐ String
☐ Potting soil
☐ Bean seeds
☐ Ice water
☐ Scissors

🏅 Water recycling

What did you learn about how water is recycled in nature?

🧠 What did we learn?

1. What two kinds of atoms combine to form water?

    a.

    b.

2. Why is water called a universal solvent?

3. What is unique about the water molecule that makes it able to dissolve so many substances?

🚀 Taking it further

1. What would happen to your body if oxygen could not be dissolved in water?

2. Is water truly a universal solvent?

3. Why is it important for mothers with nursing babies to drink lots of water?

Name _____   Date _____

🧪 Water, Water Everywhere Worksheet

Water is a vital part of life. God created water to keep plants and animals alive on earth. Because it is so important, we use water in many ways every day. See how many ways you can find water being used in your home. Fill in the chart below with every way you and your family have used water in your home in the past day.

Location	Ways water is used in this part of the home
Kitchen	
Living room	
Bathroom	
Laundry room	
Yard	
Anywhere else (Include ways your body uses water)	

Lesson 18 **Properties of Matter** 69

19 Mixtures

All mixed up

🧪 Supply list – Separating a mixture

☐ Coffee filter

☐ Orange juice

☐ Funnel

☐ Cup

🏅 Separating compounds

What methods did you determine for separating the compounds on this challenge?

🧠 What did we learn?

1. What are two differences between a compound and a mixture?

   a.

   b.

2. What is a homogeneous mixture?

3. What is a heterogeneous mixture?

4. Name three common mixtures.

   a.

   b.

   c.

🚀 Taking it further

1. If a soft metal is combined with a gas to form a hard solid that doesn't look or act like either of the original substances, is the resulting substance a mixture or a compound?

2. How might you separate the salt from the sand and water in a sample of seawater?

20 Air

What we breathe

Supply list – Importance of oxygen
☐ Small candle
☐ Jar
☐ Bottle
☐ Baking soda
☐ Vinegar

Fractional distillation
Draw a copy of the diagram on page 84 of your student book. Be sure to include the information on the different substances.

What did we learn?

1. What is likely the most important element on earth?

2. What is likely the most important compound on earth?

3. What is likely the most important mixture on earth?

4. What are the main components of air?

🚀 Taking it further

1. Why is nitrogen necessary in the air?

2. Why is oxygen necessary in air?

3. How does the composition of air show God's provision for life?

21 Milk & Cream

"Udderly" delicious

🧪 Supply list – How strong is your whipped cream?; Making butter

☐ 2 cups liquid whipping cream
☐ Jar with lid
☐ Sugar
☐ Vanilla extract
☐ Bowl
☐ Plate
☐ Electric mixer
☐ Canned spray whipped cream (made from real cream)

🧪 How strong is your whipped cream?

Which type of cream would be best to use if you want to make a dessert ahead of time and store it until your dinner guests have finished their dinner?

🏅 Supplies for Challenge – Cheese

☐ Whole milk
☐ Vinegar
☐ Pan
☐ Access to stove

🏅 Cheese

1. Were you able to make the sample of cheese? Briefly describe the process.

What did we learn?

1. Is milk an element, a compound, or a mixture?

2. What is pasteurization, and why is it done to milk?

3. What is homogenization, and why is it done to milk?

4. What is a foam?

Taking it further

1. Why does whipped cream begin to "weep"?

2. Why must cream be churned in order to make butter?

22 Solutions

Not just an answer to a problem

🧪 Supply list – Understanding solutions

☐ Roll of Life Savers® candy

☐ Rolling pin

☐ Plastic bag

☐ 3 cups

☐ Copy of "Solutions Experiments" Worksheet

🏅 Supplies for Challenge – Like dissolves like

☐ Potassium salt (potassium chloride)

☐ Table salt (sodium chloride)

☐ Sugar

☐ Dish soap

☐ Vegetable oil

☐ Baking soda

☐ 4 clear cups

☐ 2 cups (soap dissolving experiment)

☐ Copy of "Like Dissolves Like" Worksheet

🧠 What did we learn?

1. What is a solution?

2. Is a solution a homogeneous or heterogeneous mixture?

3. In a solution, what is the name for the substance being dissolved?

4. In a solution, what is the substance called in which the solute is dissolved?

5. What is solubility?

🚀 Taking it further

1. Why can more salt be dissolved in hot water than in cold water?

2. If you want sweet iced tea, would it be better to add the sugar before or after you cool the tea?

Name _____   Date_____

🧪 Solutions Experiments Worksheet

Temperature Experiment

Hypothesis: Based on what you learned in the lesson, in which cup (cold, warm, or room temperature) do you expect the Life Savers® candy to dissolve most quickly? _____

Place one Life Savers® candy in each of three cups. Add one cup of cold water to the first cup, one cup of room temperature water to the second cup, and one cup of hot tap water to the third cup. Do not stir. Time how long it takes for the Life Savers® candy to completely dissolve in each cup. Record the times below.

Temperature of the water	Time to dissolve
Cold water	
Room temperature water	
Hot tap water	

Was your hypothesis correct? _____

Surface Area Experiment

Hypothesis: Based on what you learned in the lesson, which Life Savers® candy (whole or crushed) do you predict will dissolve most quickly? _____

Place one Life Savers® candy in a plastic bag and crush it with a rolling pin. Pour all of the pieces of the crushed candy into one cup. Place a whole Life Savers® candy into a second cup. Pour one cup of hot tap water into each cup. Time how long it takes for the candy to completely dissolve in each cup. Record the times below.

Surface area of the candy	Time to dissolve
Whole candy	
Crushed candy	

Was your hypothesis correct? _____

Tongue Experiment

Hypothesis: Will a Life Savers® candy dissolve on your tongue faster if you move your tongue around or if you keep it still? _____ Try it and see if you are right.

Why do you suppose the candy dissolved faster when you moved your tongue?

Lesson 22 **Properties of Matter**

Name _____   Date _____

🎖 Like Dissolves Like Worksheet

The following substances will dissolve in water. Do you think they will dissolve in vegetable oil? Write your predictions on the chart below. Pour ¼ cup of oil in each of four clear cups. Try to dissolve ¼ teaspoon of each of the substances below in the oil. Stir each mixture, and then allow the mixtures to sit for 5 minutes. Write your observations below.

	Sodium chloride (table salt)	Potassium chloride	Baking soda	Sugar
Will it dissolve in oil? (Hypothesis)				
Did it dissolve? (Observations)				

Were your hypotheses correct? _____

Which substances did not dissolve in the oil? _____

Why do you think these substances did not dissolve? _____

_____

Soap has a different molecular structure than many other substances. Do you think it will dissolve in water? Do you think it will dissolve in oil? Write your predictions below. Stir 1 teaspoon of dish soap into ¼ cup of water and 1 teaspoon of dish soap into ¼ cup of vegetable oil. Stir each mixture then allow the mixture to stand for 5 minutes. Did the soap stay dissolved in each substance? Write your observations below.

	Water	Vegetable oil
Will soap dissolve? (Hypothesis)		
Did it dissolve? (Observations)		

Were your hypotheses correct? _____

Why do you think that soap dissolved in the substances above? _____

_____

| God's Design: Chemistry & Ecology | Properties of Matter | Day 38 | Unit 5 Lesson 23 | Name |

23 Suspensions

And we don't mean getting kicked out of school.

🧪 Supply list – Making your own suspension

☐ 1 egg
☐ Vinegar
☐ Salt
☐ Vegetable oil
☐ Dry mustard
☐ Lemon juice
☐ Paprika
☐ Small mixing bowl
☐ Electric mixer

🎖 Supplies for Challenge – Cake

☐ Cake mix
☐ Eggs
☐ Oil
☐ Water
☐ Ingredients called for in cake mix

🧠 What did we learn?

1. What is a suspension?

2. What does immiscible mean?

3. What is an emulsifier?

4. What is a colloid?

🚀 Taking it further

1. What would happen to the mayonnaise if the egg yolk were left out of the recipe?

2. How is a suspension different from a true solution?

| God's Design: Chemistry & Ecology | Properties of Matter | Day 39 | Unit 5 Lesson 24 | Name |

24 Solubility

How well does it dissolve?

🧪 Supply list – Warm and cold solutions

☐ 2 canned soft drinks (1 at room temperature and 1 refrigerated)

☐ 2 clear cups

🏅 Supplies for Challenge – Solubility of various substances

☐ Table salt

☐ Potassium salt

☐ Baking soda

☐ Sugar

☐ Spoon

☐ 4 clear cups

☐ Copy of "Solubility of Various Substances" Worksheet

🧠 What did we learn?

1. What is solubility?

2. What are the three factors that most affect solubility?

   a.

   b.

   c.

3. What is the name given to particles that come out of a saturated solution?

Lesson 24 **Properties of Matter** // 83

Taking it further

1. Why are soft drinks canned or bottled at low temperatures and high pressure?

2. Why do soft drinks eventually go flat once opened?

3. If no additional sugar has been added to a saturated solution of sugar water, what can you conclude about the temperature and/or pressure if you notice sugar beginning to settle on the bottom of the cup?

Name _____ Date _____

🎖 Solubility of Various Substances Worksheet

You will measure the solubility of four substances: table salt (sodium chloride), potassium salt (potassium chloride), baking soda, and sugar.

Hypothesis

Which of the four substances do you think you will be able to dissolve the most in a ½ cup of room temperature water?

_____

Observations

Pour ½ cup of room temperature water into each of 4 clear cups. Dissolve each substance in the water ¼ teaspoon at a time until no more will dissolve when stirred with a spoon. Record the amounts you were able to dissolve below.

	Sodium chloride (table salt)	Potassium chloride	Baking soda	Sugar
Teaspoons of solvent dissolved				

Conclusions

Which substance had the highest solubility? _____

Which substance had the lowest solubility? _____

Was your hypothesis correct? _____

Lesson 24 **Properties of Matter** 85

25 Soft Drinks

America's (second) favorite drink

🧪 Supply list – Making your own soft drink

☐ Club soda
☐ Orange juice
☐ Sugar or corn syrup
☐ Measuring spoons
☐ Vanilla extract
☐ Nutmeg
☐ Cinnamon
☐ Lemon juice
☐ Food coloring (yellow, red, blue)
☐ Baking soda
☐ Water

🎖 Supplies for Challenge – Regular or diet?

☐ Cans of regular and diet soft drinks (same brand and flavor)
☐ Large bucket

🎖 Regular or diet?

Which did you prefer? Why?

🧠 What did we learn?

1. What are the main ingredients of soft drinks?

2. What is the most popular drink in the United States? The second most popular?

3. What are the two most popular sweeteners used in soft drinks?

   a.

   b.

🚀 Taking it further

1. Why are soft drink cans warmed and dried before they are boxed?

2. Why are recipes for soft drinks considered top secret?

3. Why would the finished syrup be tested before adding the carbonation?

26 Concentration

Is your lemonade weak?

🧪 Supply list – Making ice cream

☐ Milk
☐ Ice cubes
☐ Sugar
☐ Salt
☐ Vanilla extract
☐ Quart-sized plastic zipper bag
☐ Sandwich-sized plastic zipper bag

🏅 Supplies for Challenge – Salt water

☐ Thermometer
☐ Stopwatch
☐ Sauce pan
☐ 2 cups
☐ Salt water
☐ Copy of "Salt's Effect on the Freezing and Boiling Point of Water" Worksheet
☐ Access to stove

🧠 What did we learn?

1. What is a dilute solution?

2. What is a concentrated solution?

3. How does the concentration of a solution affect its boiling point?

4. How does the concentration of a solution affect its freezing point?

Taking it further

1. Why is a quantitative observation for concentration usually more useful than a qualitative observation?

2. If a little antifreeze helps an engine run better, would it be better to add straight antifreeze to the radiator? Why or why not?

Name _____   Date _____

Salt's Effect on the Freezing & Boiling Point of Water Worksheet

Freezing Point

Time	Plain water temperature	Salt water temperature
2 minutes		
4		
6		
8		
10		
12		
14		
16		
18		
20		

At what temperature did the plain water begin to freeze? _____

At what temperature did the salt water begin to freeze? _____

What effect did the salt have on the freezing point of water? _____

_____

Boiling Point

Time	Plain water temperature	Salt water temperature
1 minute		
2		
3		
4		
5		

At what temperature did the plain water begin to boil? _____

At what temperature did the salt water begin to boil? _____

What effect did the salt have on the boiling point of water? _____

_____

27 Seawater

The world's most common solution

Supply list – Making seawater

☐ Water
☐ Straw
☐ Salt
☐ Egg
☐ Cup

What did we learn?

1. What is the most common solution on earth?

2. What are the main substances found in the ocean besides water?

3. How does salt get into the ocean?

4. Name one gas that is dissolved in the ocean water.

🚀 Taking it further

1. Why is seawater saltier than water in rivers and lakes?

2. Why is there more oxygen near the surface of the ocean than in deeper parts?

28 Water Treatment

Making it clean

🧪 Supply list – Cleaning our water

☐ Empty 2-liter plastic soft drink bottle
☐ Dirt
☐ Sand
☐ Small gravel or pebbles
☐ Alum
☐ 2 cups
☐ Dish or cup
☐ Charcoal briquettes
☐ Plastic zipper bag
☐ Cotton balls
☐ Hammer
☐ Goggles

🧠 What did we learn?

1. Why do we need water treatment plants?

2. What are the three main things that are done to water to make it clean enough for human consumption?
   a.

   b.

   c.

3. Why is it important not to dump harmful chemicals into rivers and lakes?

🚀 Taking it further

1. How is the filter you built similar to God's design for cleaning the water?

| God's Design: Chemistry & Ecology | Properties of Matter | Day 48 | Unit 6 Lesson 29 | Name |

29 Food Chemistry

You are what you eat.

🧪 Supply list – Baking soda — a vital chemical
☐ Ingredients to make your favorite cookies

🧪 Baking soda — a vital chemical

1. How do the cookies with baking soda look compared to the cookies without baking soda?

2. Are there differences in color, shape, or texture?

3. What differences do you notice in the taste? Which cookies taste better?

🏅 Supplies for Challenge – Food research
☐ Research materials on food chemicals
☐ Copy of "Food Chemicals" Worksheet

🧠 What did we learn?
1. What are the three main types of chemicals that naturally occur in food?
    a.
    b.
    c.

2. What kinds of chemicals are often added to foods?

3. Why is the kitchen a great place to look for chemicals?

🚀 Taking it further
1. If you eat a peanut butter and jelly sandwich, which part of the sandwich will be providing the most carbohydrates? The most fat? The most protein?

Name _____   Date _____

Food Chemicals Worksheet

There are a wide variety of chemicals in food. Do some research and see how many of the following questions you can answer.

1. What chemical is found in coffee and soft drinks that interferes with some people falling asleep?
   _____

2. What is the chemical name for table sugar? _____

3. What is the chemical name for baking soda? _____

4. What is the chemical name for table salt? _____

5. What chemical makes you cry when you slice onions? _____

6. What chemical gives peppers their hot flavor? _____

7. What chemical gives carrots their orange color? _____

8. What chemical gives tomatoes their red color? _____

9. What chemical gives broccoli its green color? _____

10. What chemical gives a soft drink its bubbles? _____

| God's Design: Chemistry & Ecology | Properties of Matter | Day 51 | Unit 6 Lesson 30 | Name |

30 Chemical Analysis of Food

How do I know what I'm eating?

🧪 Supply list – Analyzing your food

☐ Copy of "Chemical Analysis" Worksheets
☐ Iodine
☐ Potato or tortilla chips
☐ Apple slice
☐ Brown paper
☐ Bread
☐ Flour
☐ Vegetable oil
☐ Peanut butter
☐ Water

🎖 Supplies for Challenge – Calories

☐ Copy of "How Many Calories Did I Eat?" Worksheet
☐ Copy of "Calories Chart"

🎖 Calories

Based on what you have learned about calories, make a sample menu for dinner one night. Calculate the calories based on the chart on page 103 of the Teacher Guide.

🧠 What did we learn?

1. What are the main chemicals listed on food labels?

Lesson 30 **Properties of Matter** // 99

2. How do food manufacturers know what to put on their labels?

3. What is one way to test if a food contains oil?

4. What is an indicator?

🚀 Taking it further

1. How do you suppose indicators work?

2. Why is it important to know what chemicals are in our food?

Name _____ Date _____

🧪 Chemical Analysis—Part 1

Testing for Oils

Step 1: Place a sample of each of the foods listed on the chart below on a piece of brown paper. Write the name of the sample next to it on the paper. Allow the sample to sit for five minutes.

Step 2: While you wait, record your prediction about which samples you think will contain fat/oil in the first column of the chart.

Step 3: Wipe off any excess sample that has not evaporated.

Step 4: Hold the paper up to the light and notice which samples caused the paper to become translucent (allow light to pass through). Write your observations below.

Step 5: Check the food label (if available) to see if oil is listed in the ingredients. Write your answers below.

Step 6: Write "yes" in the last column if your test shows that the sample contains oil. Compare this with your prediction.

Sample type	Prediction	Translucent?	Oil listed in ingredients?	Oil in this sample?
Water				
Vegetable oil				
Peanut butter				
Potato or tortilla chips				
Apple slice				
Bread				
Flour				

Lesson 30 **Properties of Matter** ⚗⚗ 101

Name _____ Date _____

🧪 Chemical Analysis—Part 2

Testing for Starch

Step 1: Predict which samples you think will contain starch. Write your predictions in the first column.

Step 2: Test each of the foods listed on the chart below for the presence of starch by placing a drop of iodine on each sample.

Step 3: Record the color of the iodine after it combines with the food.

Step 4: Check the food label (if available) to see if grains such as wheat, rice, or barley are listed in the ingredients. Write your answers below.

Step 5: If the iodine turned blue or greenish, it indicates the presence of starch. Write "yes" in the last column if starch was detected. Compare this result with your prediction.

Sample type	Prediction	Color of iodine/food combination	Grain listed in the ingredients?	Starch in this sample?
Water				
Vegetable oil				
Peanut butter				
Potato or tortilla chips				
Apple slice				
Bread				
Flour				

Name _____   Date _____

Calories Chart

FOOD	SERVING SIZE	CAL
ALFALFA SEEDS, SPROUTED, RAW	1 CUP	10
ALMONDS, WHOLE	1 OZ	165
APPLE	1	60
APPLE PIE	1 PIECE	405
ARTICHOKE	1	55
ASPARAGUS	1 CUP	50
AVOCADOS	1	325
BAGELS, PLAIN	1	200
BAKED BEANS	1/4 CUP	100
BANANA	1	100
BEEF ROAST, RIB, LEAN ONLY	2.2 OZ	150
MIXED SALAD	2 CUPS	100
BISCUITS	2	100
BLACK BEANS	1 CUP	225
BLACKBERRIES, RAW	1 CUP	75
BLACK-EYED PEAS	1 CUP	190
BLT SANDWICH		500
BLUEBERRIES, RAW	1 CUP	80
BLUEBERRY PIE	1 PIECE	380
BRAZIL NUTS	1 OZ	185
CAKE OR PASTRY FLOUR, SIFTED	1 CUP	350
CARROT	1	25
CARROT CAKE, CREAM CHEESE FROSTING	1 PIECE	385
CARROTS, COOKED FROM RAW	1 CUP	70
CASHEW NUTS, DRY ROASTED, SALTED	1 OZ	165
CAULIFLOWER, COOKED	1 CUP	30
CEREAL BAR		100
CHEDDAR CHEESE	1 OZ	115
CHEESEBURGER, REGULAR		300
CHEESECAKE	1 PIECE	280
CHERRIES, SWEET, RAW	10	50
CHERRY PIE	1 PIECE	410
CHESTNUTS, ROASTED	1 CUP	350
CHICKEN, FRIED, BREAST	4.9 OZ	365
CHICKEN, FRIED, DRUMSTICK	2.5 OZ	195
CHICKPEAS, COOKED, DRAINED	1 CUP	270
CLAMS	3 OZ	65
COD	4 OZ	120
CORN GRITS, COOKED, INSTANT	1 PKT	80
CORN ON THE COB	1 EAR	155
COTTAGE CHEESE	1 CUP	235
CRACKED-WHEAT BREAD	1 SLICE	65
CREAM CHEESE	1 OZ	100
CREAM CHICKEN SOUP	1 CUP	280
CREAM PIE	1 PIECE	455
CUCUMBER, W/ PEEL	6 SLICES	5
CUSTARD PIE	1 PIECE	330
DATES	4 OZ	248
EGGS, COOKED, FRIED	1	90
EGGS, COOKED, HARD BOILED	1	75
ENCHILADA	1	235
FETA CHEESE	1 OZ	75
FIGS, DRIED	10	475
FILBERTS, (HAZELNUTS) CHOPPED	1 CUP	725
FRESH FRUIT SALAD	1 CUP	100
GARLIC BREAD	2 SLICES	200
GRAPEFRUIT JUICE	1 CUP	95
GRAPEFRUIT	1	80
GRAPES	10	40
GRAVY	2 TBSP	33
GREEN BEANS	1/2 CUP	22
GROUND BEEF, BROILED, LEAN	3 OZ	230
GROUND BEEF, BROILED, REGULAR	3 OZ	245
HALIBUT, BROILED, BUTTER, LEMON JUICE	3 OZ	140
HAM, ROASTED, LEAN	2.5 OZ	160
HONEYDEW MELON, RAW	1/10 MEL	45
ITALIAN BREAD	1 SLICE	85
JAMS AND PRESERVES	1 TBSP	55
LAMB, RIB, ROASTED, LEAN	2 OZ	130
LEMON MERINGUE PIE	1 PIECE	355
LEMONS, RAW	1	15
LETTUCE, CRISP HEAD	1 CUP	5
LIMA BEANS	1 CUP	260
MACADAMIA NUTS, SALTED	1 CUP	960
MACARONI, COOKED, FIRM	1 CUP	190
MANGOS, RAW	1	135
MILK SHAKE, THICK	12 OZ	400
MINCE PIE	1 SLICE	244
NOODLES, EGG, COOKED	1 CUP	200
OCEAN PERCH, BREADED, FRIED	1 FILLET	185
OLIVE OIL	1 TBSP	125
ONION RINGS	2	80
ONIONS, RAW, CHOPPED	1 CUP	55
ORANGE	1	50
PAPAYAS, RAW	1 CUP	65
PARSLEY, RAW	10 SPRIG	5
PEACH PIE	1 PIECE	405
PEAR	1	60
PEPPERS, SWEET, RAW	1 PEPPER	20
PICKLES, CUCUMBER, DILL	1 PICKLE	5
PINEAPPLE	1 CUP	75
PINTO BEANS	1 CUP	265
PITA BREAD	1 PITA	165
PIZZA, CHEESE	1 SLICE	290
PIZZA, CHICKEN	7' THIN	370
PLUMS	1 PLUM	15
PORK CHOP, BROIL	2.5 OZ	165
POTATO, MASHED	1/2 CUP	110
POTATO SALAD W/ MAYONNAISE	1 CUP	360
POTATOES, BAKED	1	145
POTATOES, BOILED NEW	1/4 CUP	90
PUMPKIN AND SQUASH	1 OZ	155
PUMPKIN PIE	1 PIECE	320
RADISHES, RAW	4	5
RAISINS	1 CUP	435
RASPBERRIES, RAW	1 CUP	60
RED KIDNEY BEANS	1 CUP	230
REFRIED BEANS	1 CUP	295
RICE CAKES	2	60
RICE, BROWN	1 CUP	230
RICE, FRIED	1 CUP	220
ROAST BEEF	5 OZ	250
ROAST CHICKEN	3 OZ	160
ROAST PEANUTS	2 OZ	295
ROAST TURKEY	3 OZ	140
ROLLS	1 ROLL	85
SOUR CREAM	1 CUP	495
SPAGHETTI, COOKED	1 CUP	190
SPARE RIBS	4 OZ	290
SPRING ROLL	1	120
STRAWBERRIES, RAW	1 CUP	45
SWEET AND SOUR PRAWN	1 CUP	257
SWEET POTATOES, BAKED	1	115
TACO	1	195
TOMATOES, RAW	1	25
TROUT, BROILED, W/ BUTTER, LEMON JUICE	3 OZ	175
TUNA SALAD	1 CUP	375
WAFFLES	1	205
WALNUTS, BLACK, CHOPPED	1 CUP	760
WALNUTS, ENGLISH, PIECES	1 CUP	770
WATERMELON, RAW, DICED	1 CUP	50
WHEAT FLOUR, ALL-PURPOSE, SIFTED	1 CUP	420

Name _____   Date _____

How Many Calories Did I Eat?

Food or drink	Calories per serving	Serving size	Number of servings eaten	Calories eaten

My total calories for the day = _____

31 Flavors

Chocolate or vanilla?

Supply list – Enjoying your favorite flavor; Taste and smell

☐ Life Savers® candies
☐ Instant pudding (your favorite flavor)
☐ Milk

Essential oils

Explain what essential oils are and if your family uses them. What are some of their uses?

What did we learn?

1. What two parts of your body are needed in order to fully enjoy the flavor of your food?

   a.

   b.

2. What is the difference between an herb and a spice?

3. What is the difference between a natural flavor and an artificial flavor?

Taking it further

1. Why might a cook prefer to use fresh herbs rather than dried herbs?

2. Why do you think artificial vanilla tastes different than natural vanilla even though they may have the same chemical formula?

32 Additives

What's really in your food?

🧪 Supply list – Preserving our food

☐ Apple
☐ Lemon juice
☐ Knife

🏅 Supplies for Challenge – Food additive checklist

☐ Copy of "Food Additives Checklist" Worksheet

🧠 What did we learn?

1. What is a food additive?

2. Name three different kinds of additives.
    a.
    b.
    c.

3. Why are preservatives sometimes added to foods?

4. What compound has been used as a preservative for thousands of years?

5. Why are emulsifiers sometimes added to foods?

Taking it further

1. Why are vitamins and minerals added to foods?

2. Why does homemade bread spoil faster than store-bought bread?

Name _____   Date _____

🎖 Food Additives Checklist

Emulsifiers	
Disodium phosphate	
Mono and diglycerides	
Polysorbate 60 or polysorbate 80	
Propylene glycol	
Thickeners/gelling agents	
Cellulose gum	
Xanthan gum	
Preservatives/antioxidants	
BHA and/or BHT	
Citric acid	
Potassium sorbate	
Sodium benzoate	
Flavors/flavor enhancers	
Disodium inosinate	
Disodium guanylate	
MSG (monosodium glutamate)	
Natural and/or artificial flavors	
Dough conditioner	
Calcium carbonate	
Colors	
Blue 1	
Red 40	
Yellow 5 or yellow 6	
Other additives I found:	

33 Bread

Why is it light and fluffy?

🧪 Supply list – Baking bread

- ☐ Flour
- ☐ Butter or margarine
- ☐ Yeast
- ☐ Water
- ☐ Salt
- ☐ Milk
- ☐ Sugar
- ☐ Baking pan
- ☐ Cooking oil
- ☐ Large bowl
- ☐ Small bowl

🏅 Supplies for Challenge – Homemade vs. store-bought

- ☐ Store-bought bread (with preservatives)
- ☐ 2 plastic zipper bags
- ☐ Homemade bread
- ☐ Copy of "Homemade vs. Store-bought" Worksheet

🧠 What did we learn?

1. If you want fluffy bread, what are the two most important ingredients?

2. Why is gluten important for fluffy bread?

3. Why does bread have to be baked before you eat it?

4. Why is whole wheat bread more nutritious than white bread?

Taking it further

1. What would happen if you did not put any sugar in your bread dough?

2. Can bread be made without yeast?

Name _____   Date _____

🏅 Homemade vs. Store-bought

Record your hypotheses below and then perform the experiment in your student manual.

Hypotheses

Do you expect homemade bread to dry out faster or slower than store-bought bread?
_____

Do you expect mold to grow more quickly on homemade bread or store-bought bread?
_____

Observations

Day	Homemade (in air)	Store-bought (in air)	Homemade (in bag)	Store-bought (in bag)
1				
2				
3				
4				
5				

Conclusions

Which kind of bread dried out the fastest? _____

Which kind of bread had more mold growth? _____

What effect do preservatives have on bread? _____

Which bread would you prefer to eat? Why? _____
_____

Lesson 33 **Properties of Matter**

34 Identification of Unknown Substances: Final Project

What is this, anyway?

Supply list – Identifying unknown substances

☐ Copy of "Identification of Solids" Worksheet
☐ Copy of "Identification of Liquids" Worksheet
☐ Iodine
☐ Vinegar
☐ Baking soda
☐ Water
☐ Cornstarch
☐ Oil

Identifying unknown substances

Write a paragraph explaining what you learned from each experiment so you can share the results with others.

Design your own experiment

What test did you design and who did the test? What were the results?

What did we learn?

1. What method should be used in identifying unknown substances?

2. Why should you avoid tasting unknown substances?

3. How can you test the scent of an unknown substance safely?

4. What are some physical characteristics of an unknown substance you can test at home?

5. What are some chemical characteristics you can test at home?

🚀 Taking it further

1. Why is it important for food manufacturers to test the ingredients they use and final products they produce?

2. Why is it important for water treatment facilities to test the quality of the water?

Name _____   Date _____

🧪 Identification of Solids

Carefully observe each sample and write down your observations for each substance. Then add a few drops of water to a small sample of each substance. Feel each liquid and write your observations below.

Substance	Observations of dry sample	Observations of wet sample
1		
2		
3		

Now make a hypothesis about what you think each sample might be. It's okay if you find out that you are wrong—this is a learning experience. (Younger children can ask for a list of substances to choose from.)

Substance	What I think it is
1	
2	
3	

Think of some ways that you can test your hypothesis. Here are some suggestions: Test for starch using iodine. Test for baking soda by using vinegar. Test to see if each sample will dissolve in water. Be sure to use only a small amount of your sample for each test and write your results below.

Substance	Indicates starch?	Reacts with vinegar?	Dissolves in water?
1			
2			
3			

If you are still unsure about what each substance is, you can ask your teacher. Write the correct identification for each substance below.

Sample 1 is _____

Sample 2 is _____

Sample 3 is _____

Were your hypotheses correct? _____

Lesson 34 **Properties of Matter** // 117

Name _____ Date _____

🧪 Identification of Liquids

Carefully observe each sample and write down your observations for each substance. Be sure to include how the substance looks and smells.

Substance	How it looks	How it smells
1		
2		
3		

Now make a hypothesis about what you think each sample might be. It's okay if you find out that you are wrong—this is a learning experience. (Younger children can ask for a list of substances to choose from.)

Substance	What I think it is
1	
2	
3	

Think of some ways that you can test your hypothesis. Here are some suggestions: Test how fast it evaporates by dipping your finger in the liquid then letting it evaporate. If it evaporates quickly, your finger will become cold. Test if it reacts to baking soda by dropping a few drops of each liquid on a sample of baking soda. Test its density compared to vegetable oil by adding a tablespoon of vegetable oil to each sample. If it is less dense than the oil, the oil will sink. If it is denser than the oil, the oil will float.

Substance	Evaporates quickly?	Reacts with baking soda?	Denser than oil?
1			
2			
3			

If you are still unsure about what each substance is, you can ask your teacher. Write the correct identification for each substance below.

Sample 1 is _____

Sample 2 is _____

Sample 3 is _____

Were your hypotheses correct? _____

118 ❙❙ God's Design: Chemistry & Ecology

35 Conclusion

A reliable world

Supply list – Reflect on God's creation

☐ 2 balloons
☐ Water
☐ Candle
☐ Matches or lighter

Letter to God

Write a short letter to God.

What did we learn?

1. What is the best thing you learned about matter?

Taking it further

1. What else would you like to know about matter? (Go to the library and learn about it.)

Ecological Worksheets

for Use with

Properties of Ecosystems

(*God's Design: Chemistry & Ecology*)

1 What Is an Ecosystem?

Biomes

🧪 Supply list – My backyard habitat

☐ String

☐ Yardstick/meter stick

☐ Magnifying glass

☐ Copy of "My Backyard Habitat" Worksheet

🎖 Supplies for Challenge – Ecozones

☐ Copy of "World Map"

☐ World atlas

🧠 What did we learn?

1. What is ecology?

2. What is the biosphere?

3. Give an example of something that is biotic and something that is abiotic.

4. What is flora?

5. What is fauna?

🚀 Taking it further

1. What factor has the greatest effect on the plants and animals that live in a particular ecosystem?

2. How does your habitat change throughout the day?

3. List some ways that climate affects the habitats of people.

Name _____   Date _____

🧪 My Backyard Habitat

Rope off a 1-square-yard area in your backyard and carefully examine it.

General observations

What can you see/observe while standing up?

_____
_____
_____

Closer observations

Animals or signs of animals	Plants or remains of plants	Other objects	Sounds	Weather conditions

How might animals use the things you have observed in your square? _____

_____
_____

List any other interesting observations you made about your backyard habitat.

_____
_____
_____

Lesson 1 **Properties of Ecosystems** 125

World Map

Name _____

Date _____

| God's Design: Chemistry & Ecology | Properties of Ecosystems | Day 64 | Unit 1 Lesson 2 | Name |

2 Niches

What's your job?

🧪 Supply list – An earthworm's niche; Setting up your notebook

☐ Jar
☐ Dark soil
☐ Sand
☐ Oats
☐ Earthworms
☐ Dark construction paper
☐ Tape
☐ 3-ring binder
☐ 9 dividers for the notebook

🎖 What's my niche?

Follow the instructions for the challenge. Be sure to include the following on your list:

Tree
Robin
Wolf
Mouse
Grass

🧠 What did we learn?

1. What is a niche?

2. Name two factors that determine an animal's niche.

   a.

   b.

3. What is a population?

4. What is a community?

5. What are two different kinds of niches an animal can have?
   a.

   b.

🚀 Taking it further

1. What different niches do you fill in your family and in your community?

2. How does competition for food and other resources affect the niche of a plant or animal?

3 Food Chains

Does it have links?

Supply list – Food chains & webs
☐ Drawing materials (Note: You will need to save these for Lesson 4.)

Examples:

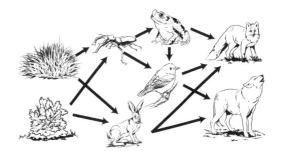

Carrying capacity
What is carrying capacity? Describe it and give your teacher an example.

What did we learn?

1. What is a food chain?

2. What is a producer?

3. What is a consumer?

4. What is a food web?

5. List two herbivores.

   a.

   b.

6. List two carnivores.

   a.

   b.

7. List two omnivores.

   a.

   b.

🚀 Taking it further

1. Is a black bear a first or second order consumer?

2. Is man an herbivore, carnivore, or omnivore?

3. Explain how a food chain shows energy flow.

| God's Design: Chemistry & Ecology | Properties of Ecosystems | Day 67 | Unit 1 Lesson 4 | Name |

4 Scavengers & Decomposers

Breaking it down

🧪 Supply list – Adding decomposers
☐ Food chain and food web pictures from lesson 3
☐ Drawing materials

Examples:

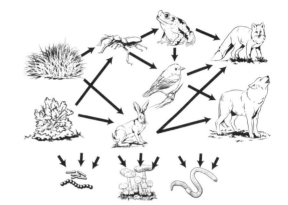

🎖 Supplies for Challenge – Population pyramids
☐ Drawing materials

🎖 Population pyramids

Draw a population pyramid for each of the food chains that you drew in lesson 3. Don't forget to put decomposers at the bottom of the pyramid. Include in your notebook.

Example:

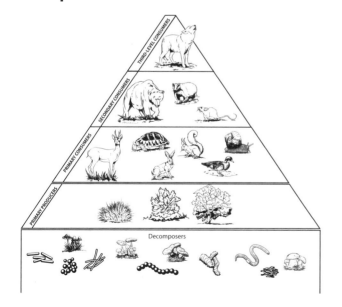

What did we learn?

1. What are organisms called that eat dead plants and animals?

2. Name two different animals that eat dead plants or animals.
   a.

   b.

3. What types of organisms are at the end of every food chain?

4. Name two common organisms responsible for decomposition.
   a.

   b.

Taking it further

1. Why is decomposition so important?

2. What physical law makes decomposition necessary?

5 Relationships among Living Things

Depending on each other

🧪 Supply list – Understanding symbiosis
☐ Copy of "Symbiosis" Worksheet
☐ Optional: Rock with lichen
☐ Magnifying glass

🏅 Supplies for Challenge – Liver flukes
☐ Research materials on liver flukes
☐ Drawing materials

🏅 Liver flukes
Make a diagram showing each step in the fluke's lifecycle. Label the relationships that occur between the fluke and the sheep, the fluke and the snail, and the fluke and the grass with the terms you have learned in this lesson. Add your diagram to your notebook.

🧠 What did we learn?
1. What is symbiosis?

2. What is mutualism?

3. What happens to each species in a parasitic relationship?

4. Which species benefits in commensalism?

5. What is competition among species?

6. What is the name of a relationship in which neither species benefits nor is harmed?

🚀 Taking it further

1. Why is competition considered harmful for both species?

2. Explain how competition could keep the species from becoming too populated.

Name _____   Date _____

🧪 Symbiosis Worksheet

Illustrate which species benefits, which is unaffected, and which is harmed in each relationship by filling in the chart with the following types of symbiosis. Some terms are used more than once.

| Mutualism | Parasitism | Commensalism | Competition | Neutralism |

+ means the species is benefited
0 means the species is neither benefited nor harmed
− means the species is harmed

Species B

Species A	+	0	−
+			
0			xxx
−		xxx	

Lesson 5 **Properties of Ecosystems** 135

6 Oxygen & Water Cycles

What comes around goes around.

🧪 Supply list – Demonstrating the water cycle
☐ Potting soil
☐ Glass jar with lid
☐ Grass or other plant
☐ Camera or drawing materials

🧪 Demonstrating the water cycle

1. What do you observe happening in the jar?

2. Does it look the same in the morning as it does in the afternoon?

🏅 Supplies for Challenge – The nitrogen cycle
☐ Research materials on the nitrogen cycle

🏅 The nitrogen cycle

Tell your teacher what the nitrogen cycle is. Be sure to make the crossword puzzle using the following words:

Ammonia
Nitrite
Nitrate
Bacteria
Legumes
Lightning

Lesson 6 **Properties of Ecosystems** // 137

What did we learn?

1. How do photosynthesis and respiration demonstrate the oxygen cycle?

2. What are the major steps in the water cycle?

Taking it further

1. Water exists in three forms: solid, liquid, and gas. What phase is the water in before and after evaporation?

2. What phase is the water in before and after condensation?

3. What phase is the water in before and after precipitation?

7 Biomes around the World

Where are they located?

Supply list – Locating biomes
☐ World atlas or online maps showing temperature and rainfall for the world as well as location of various biomes

☐ Copies of the blank "Average Rainfall," "Average Temperature," and "Biomes" world maps

Supplies for Challenge – Succession
☐ Research materials on ecological succession   ☐ Drawing materials

☐ Poster board

Succession
1. Research ecological succession on the Internet or in other sources. Then make a poster showing the same area of land in various stages of succession. Possible ideas include bare rock to forest succession after a volcanic eruption, aquatic succession, or succession after a forest fire or flood. Add this poster to your notebook.

What did we learn?
1. Where is the tropical zone located?

2. Where is the northern temperate zone located?

3. Where is the southern temperate zone located?

4. Where are the polar regions located?

Taking it further
1. Why are the polar regions generally colder than the tropical regions even though they receive many more hours of sunlight each day during the summer?

2. What correlations do you see between the temperature and rainfall maps that you made?

Name _____ Date _____

Average Rainfall

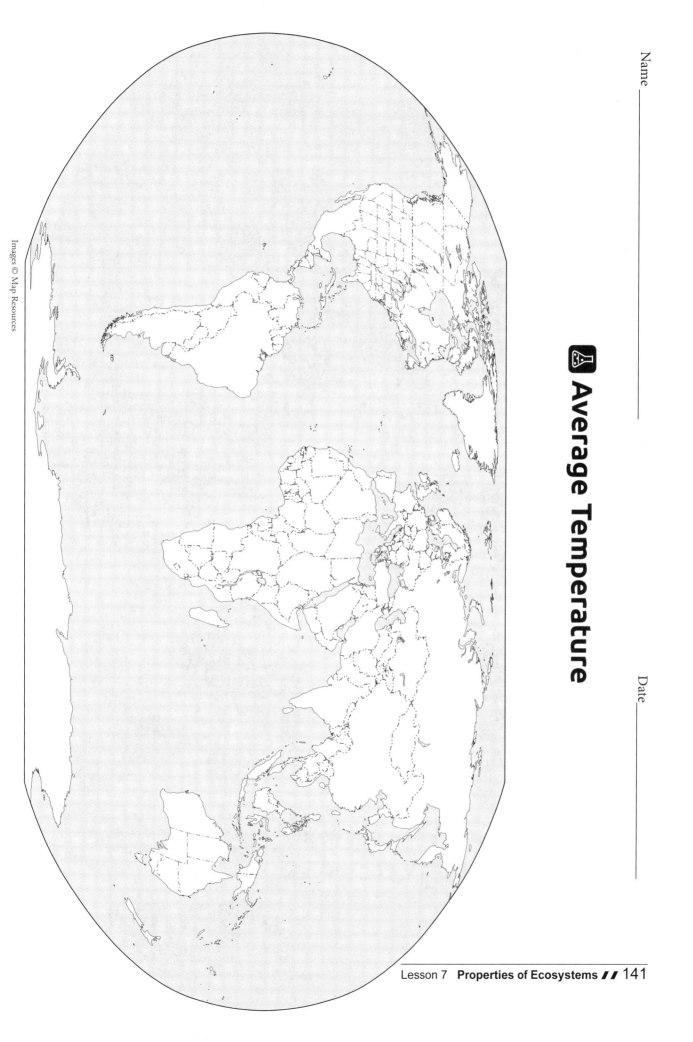

Name _____

Date _____

🧪 Biomes

| God's Design: Chemistry & Ecology | Properties of Ecosystems | Day 74 | Unit 2 Lesson 8 | Name |

8 Grasslands

Swaying in the breeze

🧪 Supply list – Grasslands worksheet; Different varieties of grass

☐ Grasses growing in a natural area

☐ Flowering plants field guide

☐ Newspaper

☐ Heavy books

☐ Cardstock or heavy paper

☐ Page protectors

☐ Copy of "Grasslands" Worksheet

🏅 Supplies for Challenge – Growing grass

☐ Grass plants

☐ Scissors

☐ Ruler

☐ Copy of "Growing Grass" Worksheet

🏅 Growing grass

1. How did cutting the grass affect its ability to grow?

2. Did one plant grow more than the others?

3. How does this experiment demonstrate God's provision for grassland animals?

What did we learn?

1. Name three characteristics of a grassland biome.
   a.

   b.

   c.

2. What are four different types of grasslands?
   a.

   b.

   c.

   d.

3. Where can each of these grasslands be found?

Taking it further

1. Why are there few trees in a grassland?

2. How do many plants survive extended periods of drought in the grassland?

3. How can grass survive when it is continually being cut down by grazing animals?

Name _____   Date _____

🧪 Grasslands Worksheet

Climate
Distinguishing features
Locations
Animals
Plants
Interesting facts

Name _____   Date _____

Growing Grass Worksheet

Day	Plant 1 height	Plant 2 height	Plant 3 height	Plant 4 height
1				
2				
3				
4				
5				
6				
7				

9 Forests

Filled with trees

🧪 Supply list – Where would I live?
☐ Copy of "Where Would I Live?" Worksheet

🎖 Supplies for Challenge – Tree anatomy
☐ Drawing materials

🎖 Tree anatomy
Draw a diagram of the layers of a tree trunk showing the bark, phloem, cambium, sapwood, and heartwood. Add this diagram to the forest section of your notebook.

🧠 What did we learn?

1. What are the major plants in a forest?

2. What are the six layers of a forest?
   a.
   b.
   c.
   d.
   e.
   f.

3. Which layer forms the roof of the forest?

4. Name three kinds of forests.

   a.

   b.

   c.

🚀 Taking it further

1. Why is the forest floor relatively dark?

2. Why is it important to study each layer of a forest?

3. How might new trees find room to grow in a mature forest?

Name _____   Date _____

🧪 Where Would I Live? Worksheet

Place each animal listed below in the layers in which you are most likely to find it. Think about the animal's habits, food, and other needs to determine where it might live. Remember, some animals may be found in more than one layer.

Bald eagle	Opossum	Hummingbird	Tree frog	Deer
Spider monkey	Black bear	Fruit bat	Ibis	Flies
Termites	Monarch butterfly	Rabbit	Lemur	Woodpecker

Layer	Animals found there
Emergent layer	
Canopy	
Understory	
Shrub layer	
Herb layer	
Floor	

Lesson 9 **Properties of Ecosystems** // 149

10 Temperate Forests

Can you see the forest for the trees?

🧪 Supply list – Tree identification; Forest worksheets
☐ Copy of "Deciduous Forest" and "Coniferous Forest" Worksheets
☐ Copy of "Tree Identification" worksheet

🏅 Supplies for Challenge – Forest jeopardy
☐ Copy of "Forest Jeopardy" Worksheet

🧠 What did we learn?

1. What are some characteristics of a deciduous forest?

2. What are some characteristics of a coniferous forest?

3. What is another name for a coniferous forest in the far north?

4. What is a deciduous tree?

5. What is a coniferous tree?

🚀 Taking it further

1. What are some ways that plants in temperate forests were designed to withstand the cold winters?

2. What are some ways that animals in temperate forests were designed to withstand the cold winters?

3. Would you expect plant material that falls to the floor of the coniferous forest to decay quickly or slowly? Why?

Name _____   Date _____

🧪 Deciduous Forest Worksheet

Climate
Distinguishing features
Locations
Animals
Plants
Interesting facts

Name _____   Date _____

🧪 Coniferous Forest Worksheet

Climate
Distinguishing features
Locations
Animals
Plants
Interesting facts

Name _____   Date _____

🧪 Tree Identification Worksheet

Look up the following trees in a tree field guide and draw a picture of what their fruits and leaves look like. If you don't have a field guide, try searching for these trees on the Internet.

A. Ginkgo biloba	B. White oak
C. Lodgepole pine	D. Sugar maple
E. Ohio buckeye	F. Common juniper

Lesson 10 **Properties of Ecosystems** // 155

Name _____   Date_____

🏅 Forest Jeopardy Worksheet

In the game *Jeopardy!*, the contestant is given an answer and must come up with the appropriate question. Here we are giving you answers that have to do with forests, and you must write an appropriate question for each one. The first one is done for you as an example.

1. Oak, maple, and beech.
   <u>What kinds of trees might you find in a deciduous forest?</u>

2. Roof of the forest.
   _____

3. Lichen, moss, and fungi.
   _____

4. Shrub layer.
   _____

5. 30–60 inches per year.
   _____

6. 12–33 inches per year.
   _____

7. Tropical and polar regions.
   _____

8. Boreal forest and Taiga.
   _____

9. Dall and big horn sheep.
   _____

10. Many lakes.
    _____

11. Duck-billed platypus.
    _____

12. Tallest trees of the forest.
    _____

156 // God's Design: Chemistry & Ecology

| God's Design: Chemistry & Ecology | Properties of Ecosystems | Day 78 | Unit 2 Lesson 11 | Name |

11 Tropical Rainforests

Growing where it's wet

🧪 Supply list – Rainforest worksheet; Researching rainforest animals
☐ Copy of "Tropical Rainforest" Summary Worksheet
☐ Research materials

Note: Choose one animal that lives in the tropical rainforest and write a report on it to include in your notebook. Include pictures of your animal. Find answers to the questions on page 180 of your student book.

🎖 Rainforest products
Note: This optional challenge can be completed using tape or glue, blank paper, old magazines, scrap paper, crayons, or colored pencils.

🧠 What did we learn?

1. List some ways in which a tropical rainforest is different from a temperate forest.

2. Where are the rainforests located?

3. What is an arboreal animal?

4. What is an epiphyte?

5. Name at least one epiphyte.

🚀 Taking it further

1. Do you think that dead materials would decay slowly or quickly on the floor of the rainforest? Why?

2. If you transplanted trees such as orange, cacao, or papaya trees to a deciduous forest, would you expect them to survive? Why or why not?

3. Which animals are you most likely to see if you are taking a walk through the tropical rainforest?

Name _____   Date _____

🧪 Tropical Rainforest Summary Worksheet

Climate
Distinguishing features
Locations
Animals
Plants
Interesting facts

| God's Design: Chemistry & Ecology | Properties of Ecosystems | Day 81 | Unit 3 Lesson 12 | Name |

12 The Ocean

Marine ecosystem

🧪 Supply list – Ocean worksheet; Currents distribute nutrients
☐ Copy of "Ocean" Summary Worksheet
☐ Shallow pan
☐ Water
☐ Food coloring

🏅 Supplies for Challenge – Understanding bioluminescence
☐ Drawing materials

🧠 What did we learn?

1. How much of the earth is covered with water?

2. How much of the surface water of the world is in the ocean?

3. How many oceans are there?

4. What are the three zones that the ocean can be divided into?
   a.
   b.
   c.

5. What are the three major groups of living organisms in the ocean?
   a.
   b.
   c.

Taking it Further

1. What might happen in the ocean if the currents stopped flowing?

2. Why do most animals in the ocean live in the euphotic zone?

3. Why might the aphotic zone occur at a shallower depth than 660 feet (200 m) in some areas?

Name _____   Date _____

🧪 Ocean Summary Worksheet

Climate

Distinguishing features

Locations

Animals

Plants

Interesting facts

13 Coral Reefs

Underwater wonderlands

🧪 Supply list – Coral reef worksheet; Coral model
☐ Copy of "Coral Reef" Summary Worksheet
☐ Modeling clay

🎖 Coral bleaching
Discuss coral bleaching with your teacher — what it is and what causes it.

🧠 What did we learn?

1. Where will you find coral reefs?

2. What is a coral reef made from?

3. Where do corals get most of their energy?

4. What are the three main types of coral reefs?
   a.
   b.
   c.

5. What are some of the animals that live in a coral reef besides corals?

🚀 Taking it Further

1. Why are coral reefs found in water that is usually less than 150 feet (45 m) deep?

2. Why do corals grow best in swift water?

Name _____   Date _____

🧪 Coral Reef Summary Worksheet

Climate
Distinguishing features
Locations
Animals
Plants
Interesting facts

14 Beaches

Take a walk on the sand

🧪 Supply list – Beach worksheet; Making sand

☐ Copy of "Beach" Summary Worksheet
☐ Rocks
☐ Seashells
☐ Plastic zipper bag
☐ Hammer
☐ Safety goggles
☐ Towel
☐ Sand
☐ Magnifying glass

🎖 Supplies for Challenge – A dune system

☐ Drawing materials

🧠 What did we learn?

1. What is a beach?

2. What are the two main kinds of beaches?
   a.

   b.

3. What is the name of the area of land that is covered at high tide and uncovered at low tide?

4. What are some animals you are likely to see in a beach ecosystem?

Taking it further

1. Why might you find different plants and animals on a rocky beach from those on a sandy beach?

2. How is new sand formed?

3. Explain how a beach can be in dynamic equilibrium.

Name _____   Date _____

🧪 Beach Summary Worksheet

Climate

Distinguishing features

Locations

Animals

Plants

Interesting facts

15 Estuaries

Where fresh and salty meet

🧪 Supply list – Mixing fresh & saltwater; Estuary worksheet

☐ Copy of "Estuary" Summary Worksheet

☐ 4 clear cups

☐ Water

☐ Salt

☐ Eye dropper

☐ Marker

☐ Green and blue food coloring

🏅 Watersheds

Research what watershed you are part of and note it here:

🧠 What did we learn?

1. What is an estuary?

2. Name three types of estuaries.

   a.

   b.

   c.

3. What are some plants you might find in an estuary?

4. Name several animals that you might find in an estuary.

🚀 Taking it further

1. Why is an estuary a very productive ecosystem?

2. How do mangrove trees help coral reefs?

3. Why is the salt level in the water constantly changing in an estuary?

4. Why might you find different estuary animals in the same location at different times of the year?

Name _____ Date _____

🧪 Estuary Summary Worksheet

Climate
Distinguishing features
Locations
Animals
Plants
Interesting facts

| God's Design: Chemistry & Ecology | Properties of Ecosystems | Day 86 | Unit 3 Lesson 16 | Name |

16 Lakes & Ponds

It's fresh

🧪 Supply list – Lakes & ponds worksheet; Freezing fresh & saltwater

☐ Copy of "Lakes & Ponds" Summary Worksheet

☐ 2 clear cups

☐ Water

☐ Salt

☐ Thermometer

☐ Marker

☐ Copy of "Watching Water Freeze" Worksheet

🏅 Supplies for Challenge – The Great Lakes

☐ Copy of "Great Lakes Fact Sheet"

🧠 What did we learn?

1. What is a lake?

2. What is a pond?

3. What are two ways that lakes were formed in the past?
    a.

    b.

4. What is an overturn?

5. What is an algae bloom?

🚀 Taking it further

1. Why is overturn important to lake ecosystems?

2. Why does an algae bloom often occur in a lake in the spring?

3. In which lake zone would you expect to find most small creatures like rotifers?

4. What would happen to fish during the winter if ice did not float?

Name _____   Date _____

🧪 Lakes & Ponds Summary Worksheet

Climate
Distinguishing features
Locations
Animals
Plants
Interesting facts

Name _____   Date _____

🧪 Watching Water Freeze Worksheet

Time	Freshwater temperature	Freshwater observations	Saltwater temperature	Saltwater observations
0 min.				
10 min.				
20 min.				
30 min.				
40 min.				
50 min.				
60 min.				

Name _____ Date _____

Great Lakes Fact Sheet

Feature	Lake Superior	Lake Michigan	Lake Huron	Lake Erie	Lake Ontario
Average depth					
Maximum depth					
Volume					
Major cities that border it					

17 Rivers & Streams

Flowing water

🧪 Supply list – Rivers & streams worksheet; Rivers of the world

☐ Copy of "Rivers & Streams" Summary Worksheet
☐ Copy of "Rivers of the World" Map
☐ World atlas

🏅 Supplies for Challenge – River facts

☐ Copy of "River Facts Sheet"

🧠 What did we learn?

1. What is a river?

2. Where does most of the energy for a river ecosystem come from?

3. Name some plants you might find in a river ecosystem.

4. What is a tributary?

5. What is the riparian zone?

Taking it further

1. Why do fewer plants grow in the water of a river than in a lake or ocean?

2. Would you expect a river to be larger at a higher elevation or a lower elevation?

3. Do rivers move faster over steep ground or in relatively flat areas?

4. Would you expect water to cause more erosion in a steep area or in a relatively flat area?

Name _____   Date _____

🧪 Rivers & Streams Summary Worksheet

Climate
Distinguishing features
Locations
Animals
Plants
Interesting facts

Rivers of the World

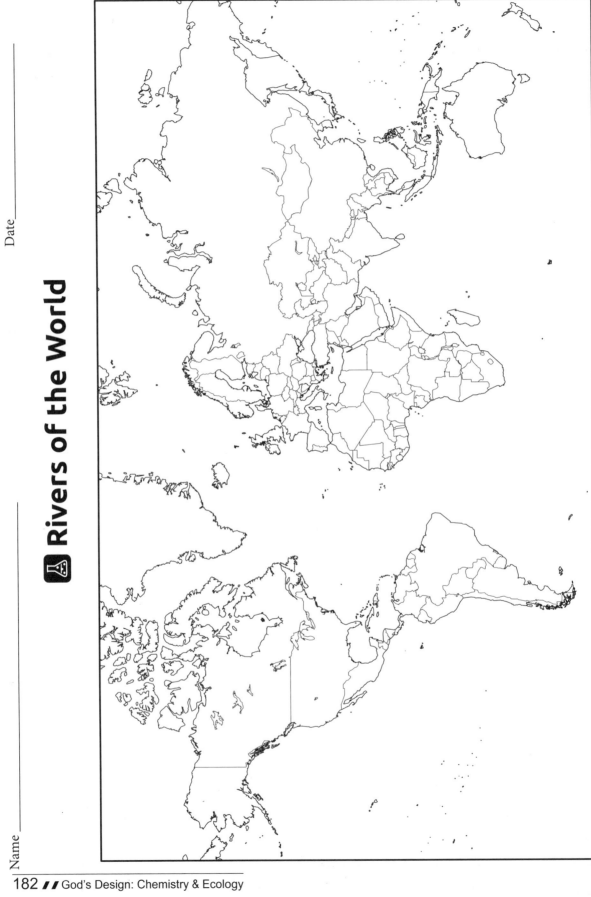

Name _____ Date _____

🎖 River Facts Sheet

River	Length	Size of river basin	Discharge at mouth	Countries or states it flows through	Major tributaries
Amazon					
Congo					
Nile					
Mississippi					
Yangtze					
Rio De La Plata					
Hwang Ho/ Yellow					
Orinoco					
Yukon					
Volga					

| God's Design: Chemistry & Ecology | Properties of Ecosystems | Day 91 | Unit 4 Lesson 18 | Name |

18 Tundra

Is it frozen?

🧪 Supply list – Tundra worksheet; Animals changing colors; Two layers of fur

☐ Copy of "Tundra" Summary Worksheet
☐ Small box
☐ White cotton balls
☐ Large bowl
☐ Ice
☐ White tissue paper or white quilt batting
☐ Photos of Arctic animals with white fur or feathers
☐ Two pairs of gloves (one pair must fit inside the other; for example, one could be cotton gardening gloves and the other could be leather work gloves)

🏅 Polar bears

Draw an image of a polar bear or give your teacher a short oral report on this beautiful animal.

🧠 What did we learn?

1. Where is most tundra located?

2. What is permafrost?

3. What kind of plants grow in the tundra?

4. What are some animals you might find in the tundra?

5. How much precipitation does the tundra receive?

🚀 Taking it further

1. Why do many animals in the tundra have white fur or feathers?

2. Why do many animals and plants have an accelerated life cycle in the tundra?

3. Why do you think the temperatures are so cool in the summer when there is often 24 hours of sunshine?

Name _____   Date _____

🧪 Tundra Summary Worksheet

Climate

Distinguishing features

Locations

Animals

Plants

Interesting facts

19 Deserts

Sand and more sand

Supply list – Desert worksheet; Storing water

☐ Copy of "Desert" Summary Worksheet

☐ Thin plastic bag (such as a produce bag)

What did we learn?

1. What is a desert ecosystem?

2. How is a cold desert different from a hot desert?

3. What are some plants you would expect to find in the desert?

4. What are some animals you would expect to find in the desert?

5. What is the difference between a Bactrian camel and a dromedary camel?

🚀 Taking it further

1. In what ways are plants well suited for the desert environment?

2. In what ways are animals well suited for the desert environment?

3. Why does rain often cause flash flooding in the desert?

4. What are some dangers you may face in the desert?

5. Why do salt flats often form in the desert?

6. Would you expect to find more salt flats in a cold desert or a hot desert?

Name _____   Date _____

🧪 Desert Summary Worksheet

Climate

Distinguishing features

Locations

Animals

Plants

Interesting facts

| God's Design: Chemistry & Ecology | Properties of Ecosystems | Day 94 | Unit 4 Lesson 20 | Name |

20 Oases

A refreshing spot

🧪 Supply list – Transpiration; Oasis worksheet

☐ Copy of "Oasis" Summary Worksheet

☐ Several plant leaves

☐ Plastic zipper bag

🎖 Supplies for Challenge – Products of the desert

☐ Research materials on desert products

☐ Poster board

☐ Drawing materials

🎖 Products of the desert

Do a little investigation of your own. Find out how the products of the desert are being used. Then, create a poster or report about desert products to include in your notebook.

🧠 What did we learn?

1. What is an oasis?

2. What kinds of plants grow in an oasis?

3. What kinds of animals live in an oasis that don't usually live in a desert?

🚀 Taking it further

1. Why is it often cooler in an oasis than in a desert?

2. Why are oases important for trade routes?

3. How might a man-made oasis change the ecosystem in a desert?

Name _____ Date _____

🧪 Oasis Summary Worksheet

Climate

Distinguishing features

Locations

Animals

Plants

Interesting facts

21 Mountains

Purple mountain majesties

🧪 Supply list – Mountain worksheet; Modeling a mountain

☐ Copy of "Mountain" Summary Worksheet
☐ Art supplies
☐ Newspaper
☐ Paint
☐ Leaves
☐ Twigs
☐ Grass
☐ Small flowers
☐ Cotton balls

🏅 Supplies for Challenge – The Himalayas

☐ Research materials on the Himalayas

🧠 What did we learn?

1. What ecosystems are you likely to encounter on mountains in temperate zones?

2. What ecosystems are you likely to encounter on mountains in tropical zones?

3. What is timberline?

4. What is snow line?

Taking it further

1. Why do the ecosystems change as you gain altitude on a mountain?

2. Why don't you find every ecosystem on every mountain?

3. What other ecosystems are you likely to find on mountains that were not listed in this lesson?

4. How have glaciers influenced the shapes of mountains?

5. Why is there less oxygen as you gain altitude?

Name _____   Date _____

🧪 Mountain Summary Worksheet

Climate
Distinguishing features
Locations
Animals
Plants
Interesting facts

Lesson 21 **Properties of Ecosystems** 197

22 Chaparral

The Mediterranean climate

🧪 Supply list – Chaparral worksheet; Fire in the chaparral
☐ Copy of "Chaparral" Summary Worksheet
☐ Pictures of the chaparral

🎖 Fire germination
Discuss fire germination with your teacher.

🧠 What did we learn?

1. What is a chaparral ecosystem?

2. What are two other names for chaparral?
   a.

   b.

3. Name some plants you might find in the chaparral.

4. Name some animals you might find in the chaparral.

5. What animal might you find in the Australian chaparral that you would not find in the American chaparral?

🚀 Taking it further

1. What conditions make fire likely in the chaparral?

2. How are plants in the chaparral specially designed for fire?

3. Should people try to put out fires that naturally occur in the chaparral?

Name _____   Date _____

🧪 Chaparral Summary Worksheet

Climate
Distinguishing features
Locations
Animals
Plants
Interesting facts

23 Caves

Are they just holes in the ground?

🧪 Supply list – Cave worksheet; Plants in caves?
☐ Copy of "Cave" Summary Worksheet
☐ Houseplant
☐ Box

🏅 Supplies for Challenge – Bats
☐ Drawing materials

🧠 What did we learn?

1. What is a cave?

2. What kinds of plants will you find in a cave ecosystem?

3. What are the three categories of animals in a cave ecosystem?
   a.
   b.
   c.

4. Explain the different habits of each category of cave animal.

5. What is the main source of nutrients in a cave ecosystem?

🚀 Taking it further

1. Why is a cave considered a low energy ecosystem?

2. Why can a rise in temperature inside a cave threaten the ecosystem?

3. Which sense is least useful in a cave?

4. Which senses are most useful in a cave?

Name _____   Date _____

🧪 Cave Summary Worksheet

Climate
Distinguishing features
Locations
Animals
Plants
Interesting facts

24 Seasonal Behaviors

It happens every year.

🧪 Supply list – Monarch butterflies

☐ Research materials on monarch butterflies

☐ Paper

☐ Colored pencils

🧠 What did we learn?

1. What is hibernation?

2. What is estivation?

3. What is migration?

4. List three different kinds of animals that migrate.

    a.

    b.

    c.

5. What is the most likely trigger for seasonal behaviors?

Taking it further

1. How can animals know where they are supposed to go when they migrate if they have never been there before?

2. How do animals navigate while migrating?

3. Why might a group of animals move from one location to another, other than for their annual migration?

4. If you see a monarch butterfly in the fall and then see another one in the spring, how likely is it that you are seeing the same butterfly?

25 Animal Defenses

A matter of protection

🧪 Supply list – Animal defenses
☐ Card stock or tag board
☐ Drawing materials
☐ Pictures of animals

🎖 Plant defenses
Talk to your teacher about plant defenses and how they connect to the Fall of Man in the Garden of Eden.

🧠 What did we learn?

1. What are three main ways that animals try to defend themselves?

   a.

   b.

   c.

2. List three ways that animals can trick their enemies into leaving them alone.

   a.

   b.

   c.

3. How do some eels protect themselves?

🚀 Taking it further

1. Why do you think animals prefer to run away or frighten off enemies rather than fight?

2. Why do many animals prefer trickery to running away?

3. How might a defense also serve as an attack method?

26 Adaptation

Fitting in

Supply list – Design worksheet
☐ Copy of "How Was I Designed?" Worksheet

Darwin's finches
What are your thoughts about Darwin's finches?

What did we learn?
1. What is adaptation?

2. Are all helpful characteristics a result of a change in the organism?

3. What process causes different species to develop among the same kind of animal or plant?

Taking it further
1. How does natural selection work?

2. Does natural selection require millions of years to develop distinct populations?

3. Does natural selection require genetic mutation?

Name _____   Date_____

🧪 How Was I Designed? Worksheet

Indicate the ways that these organisms are adapted to their environment.

Organism	Design features
Jack rabbit	
Woodpecker	
Orchid	
Honey bee	
Cactus	
Brown bat	
Oak tree	
Prairie grass	
Barn owl	
Chameleon	

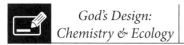

27 Balance of Nature

Keeping it working

Supply list – Population diagrams; Growing your own populations

☐ Drawing materials
☐ Cooking pot
☐ Grass
☐ Distilled water
☐ Jar

☐ Microscope
☐ Eyedropper
☐ Microscope slides and cover slips
☐ Copy of "Growing an Ecosystem" Worksheet
☐ pH testing paper

What did we learn?

1. What is meant by the balance of nature?

2. Name two ways that the balance of nature is maintained in an ecosystem.
   a.
   b.

3. What are two ways that animals use to stake out their territory?
   a.
   b.

4. What happens if a male cannot find a territory to defend?

🚀 Taking it further

1. What would be the likely effect on the ecosystem if a prairie dog colony was devastated by the plague?

2. What would happen if animals did not respect each others' territories?

3. How does the oxygen cycle demonstrate the balance of nature?

4. Which methods of population control may have been present originally, and which have developed since the Fall?

Name _____   Date _____

🧪 Growing an Ecosystem Worksheet

Day	pH	Observations
1		
2		
3		
4		
5		
6		
7		

| God's Design: Chemistry & Ecology | Properties of Ecosystems | Day 108 | Unit 6 Lesson 28 | Name |

28 Man's Impact on the Environment

Where do we fit in?

Supply list – Recording my impact
☐ Copy of "How I Impact Nature" Worksheet

A biblical view of ecology
Optional: Create a poster or page for your notebook that explains the biblical view of ecology.

What did we learn?

1. What are some ways that farmers impact ecosystems?

2. What are some ways that farmers and ranchers have changed their practices to be more friendly to the environment?

3. What are some ways that industry impacts ecosystems?

Taking it further

1. What are some ways that people can minimize their impact on nature?

2. How can hunting licenses positively affect man's impact on ecosystems?

Lesson 28 **Properties of Ecosystems** 217

Name _____   Date _____

🧪 How I Impact Nature Worksheet

Activity	Effect on ecosystems
Foods I eat	
Transportation I use	
Clothes I wear	
Recreational activities	

29 Endangered Species

Are they disappearing?

🧪 Supply list – Researching endangered species
☐ Research materials on endangered species

🏅 Wildlife management
Talk to your teacher about wildlife management.

🧠 What did we learn?

1. Name two possible natural causes of extinction of a species.

    a.

    b.

2. Name three possible man-made causes of extinction of a species.

    a.

    b.

    c.

3. Name three things people are doing to help endangered species.

    a.

    b.

    c.

🚀 Taking it further

1. Why might people overhunt a particular animal?

2. Can people use the land without harming endangered species?

| God's Design: Chemistry & Ecology | Properties of Ecosystems | Day 112 | Unit 6 Lesson 30 | Name |

30 Pollution

What happened to clean air?

🧪 Supply list – Measuring your trash

☐ Rubber gloves

☐ Newspaper

☐ Bathroom scale

☐ One week of family trash

☐ Copy of "Our Family's Trash" Worksheet

🏅 Supplies for Challenge – Blocking UV radiation

☐ Clear plastic sheet protector

☐ Sunscreen lotion

☐ Newspaper

☐ Modeling clay

🧠 What did we learn?

1. What is pollution?

2. What are some natural sources of pollution?

3. What are some sources of man-made pollution?

4. What are three major areas of the environment that can become polluted?
   a.

   b.

   c.

Taking it further

1. What are some ways that people can reduce water pollution?

2. What are some ways that people can reduce air pollution?

3. What are some ways that people can reduce land pollution?

4. Do you think that water, air, and land are cleaner or dirtier today than they were 40 years ago?

Name_____   Date_____

🗑️ Our Family's Trash Worksheet

	One week		One Year	
Type of trash	Amount sent to dump	Amount recycled	Amount sent to dump	Amount recycled
Metal				
Glass				
Paper				
Plastic				
Other				

Ways we can reduce our trash:

_____

_____

_____

_____

Lesson 30 **Properties of Ecosystems**

| God's Design: Chemistry & Ecology | Properties of Ecosystems | Day 113 | Unit 6 Lesson 31 | Name |

31 Acid Rain

Does it burn?

🧪 Supply list – Effects of acid rain

☐ 2 identical houseplants

☐ 2 spray bottles

☐ Vinegar

☐ Water

☐ Copy of "Acid Rain" Worksheet

🏅 Alternative energy sources

Optional: Make a presentation of what you learned about alternative energy sources and share it with your teacher. Include what you learned in your notebook. Be creative in your presentation.

🧠 What did we learn?

1. Why is rain naturally slightly acidic?

2. What is acid rain?

3. What are the main causes of acid rain?

4. What is buffering capacity?

🚀 Taking it Further

1. What are some ways to help reduce acid rain?

2. If the buffering capacity were the same, would you expect acid rain to be more of a problem or less of a problem in areas with high population densities? Why?

Name _____ Date _____

🧪 Acid Rain Worksheet

Day	Plant with water	Plant with acid rain
1		
2		
3		
4		
5		
6		
7		

Which plant looks healthier after 7 days? _____

32 Global Warming

Is it really heating up?

🧪 Supply list – The greenhouse effect

☐ 2 thermometers

☐ Glass jar with a lid

☐ Copy of "The Greenhouse Effect" Worksheet

🏅 Supplies for Challenge – More alternative energy sources; Carbon dioxide: a greenhouse gas

☐ Copy of "Carbon Dioxide: A Greenhouse Gas" Worksheet

☐ 2 1-gallon plastic zipper bags

☐ Vinegar

☐ Poster board

☐ Drawing materials

☐ Baking soda

☐ Two thermometers

☐ 2-liter soda bottle

🧠 What did we learn?

1. What is the greenhouse effect?

2. Why is the greenhouse effect important on earth?

3. What is global warming?

4. What do many scientists claim are the two main causes of global warming?

🚀 Taking it further

1. What are some ways that people might reduce the amount of carbon dioxide they are putting into the atmosphere?

2. Why is it inappropriate to panic about global warming?

Name _____   Date _____

🧪 The Greenhouse Effect Worksheet

Time (minutes)	Temperature outside the jar	Temperature inside the jar
0		
5		
10		
15		
20		
25		
30		

Conclusion

How did the glass jar affect the temperature inside the jar?

_____

_____

_____

How is this similar to the greenhouse effect in the environment?

_____

_____

_____

Name _____ Date _____

🎗 Carbon Dioxide: A Greenhouse Gas Worksheet

Do you think a bag containing air or a bag containing carbon dioxide will have a higher temperature after sitting in the sun for 15 minutes? Write your guess below.

Hypothesis

_____

Data

	Air	Carbon Dioxide
Initial Temperature		
Final Temperature		

Which bag trapped the most heat? _____

Does air contain greenhouse gases? How do you know? _____

_____

Is carbon dioxide a greenhouse gas? How do you know? _____

_____

33 What Can You Do?

How can I help?

Supply list – Making a plan
☐ 2 copies of "3 R's of Conservation" Worksheet

Plastic recycling
Do you recycle?

What did we learn?

1. What are the three R's of conservation?

    a.

    b.

    c.

2. List two ways you plan to do each of these things.

    a.

    b.

Taking it further

1. Why is it important to be concerned about how humans impact the environment?

Name _____ Date _____

3 R's of Conservation Worksheet
My Family's Plan

Reduce	Reuse	Recycle

Name _____ Date _____

🧪 3 R's of Conservation Worksheet
My Family's Plan

Reduce	Reuse	Recycle

| God's Design: Chemistry & Ecology | Properties of Ecosystems | Day 117 | Unit 6 Lesson 34 | Name |

34 Reviewing Ecosystems: Final Project

Final Project Supply list
☐ Supplies will vary depending on project

| God's Design: Chemistry & Ecology | Properties of Ecosystems | Day 119 | Unit 6 Lesson 35 | Name |

35 Conclusion

Appreciating our beautiful but cursed world

Supply list – A poem
☐ Bible
☐ Writing materials

Atomic & Molecular Worksheets

for Use with

Properties of Atoms & Molecules

(*God's Design: Chemistry & Ecology*)

| God's Design: Chemistry & Ecology | Properties of Atoms & Molecules | Day 122 | Unit 1 Lesson 1 | Name |

1 Introduction to Chemistry

The study of matter and molecules

Supply list – Chemistry is fun

☐ Drinking cup

☐ Baking soda

☐ Vinegar

Supplies for Challenge – Soda fountain

☐ 2-liter bottle of diet soda

☐ Mentos® mints

☐ Toothpick

☐ Tape

☐ Heavy paper

What did we learn?

1. What is matter?

2. Does air have mass?

3. What do chemists study?

🚀 Taking it further

1. Would you expect to see the same reaction each time you combine baking soda and vinegar?

| God's Design: Chemistry & Ecology | Properties of Atoms & Molecules | Day 123 | Unit 1 Lesson 2 | Name |

2 Atoms

Basic building blocks

Supply list – Atomic models
☐ Copy of "Atomic Models" Worksheet
☐ Colored pencils

Supplies for Challenge – Energy levels
☐ Copy of "Energy Levels" Worksheet

What did we learn?

1. What is an atom?

2. What are the three parts of an atom?
   a.

   b.

   c.

3. What electrical charge does each part of the atom have?

4. What is the nucleus of an atom?

5. What part of the atom determines what type of atom it is?

6. What is a valence electron?

🚀 Taking it further

1. Why is it necessary to use a model to show what an atom is like?

2. On your worksheet, you colored neutrons blue and protons red. Are neutrons actually blue and protons actually red in a real atom?

Name _____   Date _____

🧪 Atomic Models Worksheet

Color the protons in each atom red, the neutrons blue, and the electrons gray.

Hydrogen

Hydrogen is the simplest atom. It has one proton, zero neutrons, and one electron.

Helium has two protons, two neutrons, and two electrons.

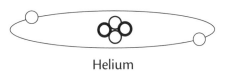

Helium

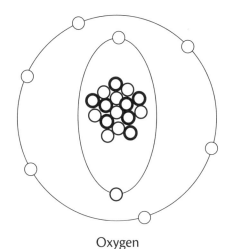

Oxygen

Oxygen has eight protons, eight neutrons, and eight electrons.

Potassium is a bit more complex with 19 protons, 20 neutrons, and 19 electrons.

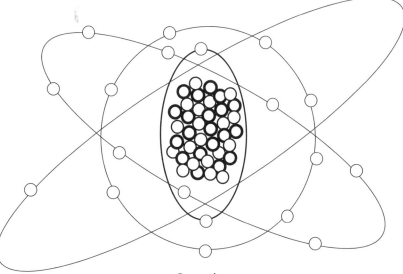

Potassium

Lesson 2 **Properties of Atoms & Molecules** ⁄⁄ 245

Name _____   Date _____

Energy Levels Worksheet

For each element listed below, tell how many energy levels contain electrons and how many electrons are in each of those levels. Use the periodic table of the elements in your student manual to help you find the answers.

Element	Energy levels	Electrons in level 1	Electrons in level 2	Electrons in level 3	Electrons in level 4	Electrons in level 5	Electrons in level 6
He Helium							
Be Beryllium							
Al Aluminum							
Cl Chlorine							
Fe Iron							
Kr Krypton							
Ag Silver							
Au Gold							

| God's Design: Chemistry & Ecology | Properties of Atoms & Molecules | Day 124 | Unit 1 Lesson 3 | Name |

3 Atomic Mass

How big is an atom?

🧪 Supply list – Learning about atoms
☐ Copy of "Learning About Atoms" Worksheet

🏅 Supplies for Challenge – Isotopes
☐ Copy of "Understanding Atoms" Worksheet

🧠 What did we learn?

1. What are the three particles that make up an atom?
   a.

   b.

   c.

2. What is the atomic number of an atom?

3. What is the atomic mass of an atom?

4. How can you determine the number of electrons, protons, and neutrons in an atom if you are given the atomic number and atomic mass?

Lesson 3 Properties of Atoms & Molecules 247

Taking it further

1. What does a hydrogen atom become if it loses its electron?

2. Why are electrons ignored when calculating an element's mass?

Name _____ Date _____

Learning About Atoms Worksheet

Complete the following chart. You can figure out the numbers of protons, electrons, and neutrons from the atomic number and atomic mass.

Element	Atomic number	Atomic mass	# of protons	# of electrons	# of neutrons
Hydrogen	1	1			
Helium	2	4			
Oxygen	8	16			
Fluorine	9	19			
Chromium	24	52			

Lesson 3 **Properties of Atoms & Molecules**

Name _____   Date _____

🎖 Understanding Atoms Worksheet

Use a periodic table to help fill in the chart below. The symbol for each element is in bold letters near the top of each square. The atomic number (which is equal to the number of protons) is above the symbol. The atomic mass is below the symbol.

Element name	Symbol	Atomic number	Atomic mass	# of protons	# of electrons	Most common # of neutrons
Hydrogen						
Oxygen						
Boron						
Gold						
Silver						
Uranium						
Potassium						
Chlorine						
Neon						
Einsteinium						

| God's Design: Chemistry & Ecology | Properties of Atoms & Molecules | Day 127 | Unit 1 Lesson 4 | Name |

4 Molecules

Putting atoms together

Supply list – Understanding molecules
☐ Copy of "What Am I?" Worksheet

Supplies for Challenge – Molecule puzzle pieces
☐ Copy of "Molecule Puzzle Pieces"
☐ Scissors

What did we learn?

1. What is a molecule?

2. What is a diatomic molecule?

3. What is a compound?

Taking it further

1. What is the most important factor in determining if two atoms will bond with each other?

2. Table salt is a compound formed from sodium and chlorine. Would you expect sodium atoms and chlorine atoms to taste salty? Why or why not?

Name _____   Date _____

🧪 What Am I? Worksheet

Next to each of the substances below, write whether it is an element, a diatomic molecule, or a compound. Review these terms in the lesson if you need to.

1. Gold (Au) – yellow metal, number 79 on the periodic table _____

2. Ammonia (NH_3) – liquid formed from 1 nitrogen and 3 hydrogen atoms _____

3. Oxygen (O_2) – gas formed from 2 oxygen atoms _____

4. Nitrogen (N_2) – gas formed from 2 nitrogen atoms _____

5. Silver (Ag) – shiny metal, number 47 on the periodic table _____

6. Salt (NaCl) – solid formed from 1 sodium and 1 chlorine atom _____

7. Sucrose ($C_{12}H_{22}O_{11}$) – solid formed from 12 carbon atoms, 22 hydrogen atoms, and 11 oxygen atoms
   _____

8. Helium (He) – gas, number 2 on the periodic table _____

9. Water (H_2O) – liquid formed from 2 hydrogen atoms and 1 oxygen atom _____

10. Baking soda ($NaHCO_3$) – powder composed of 1 sodium atom, 1 hydrogen atom, 1 carbon atom, and 3 oxygen atoms
    _____

Name _____ Date _____

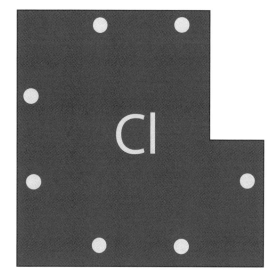
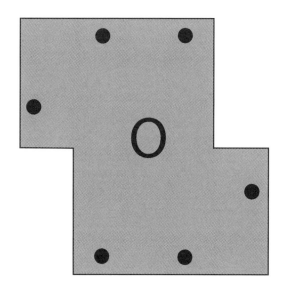

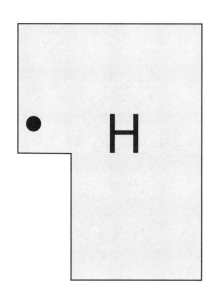
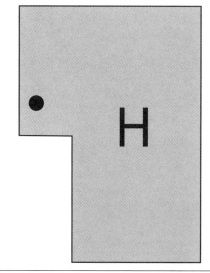

Lesson 4 **Properties of Atoms & Molecules** // 253

Periodic Table of the Elements

Organizing the elements

🧪 Supply list – Using the periodic table
☐ Copy of "Learning About the Elements" Worksheet

🎖 Synthetic elements
Do a short oral report on synthetic elements or research one that is used commercially.

🧠 What did we learn?

1. How many valence electrons do the elements in each column have?

2. What four pieces of information are included for each element in any periodic table of the elements?
   a.
   b.
   c.
   d.

3. What do all elements in a column on the periodic table have in common?

4. What do all elements in a row on the periodic table have in common?

🚀 Taking it further

1. Atoms are stable when they have eight electrons in their outermost energy level. Therefore, elements from column IA will react easily with elements from which column?

2. Elements from column IIA will react easily with elements from which column?

Name _____   Date _____

🧪 Learning About the Elements Worksheet

Use the periodic table of the elements in the student manual to answer the following questions.

1. What is the symbol for calcium? _____

2. What is the symbol for silver? _____

3. What is the atomic number for copper? _____

4. What is the atomic mass for rutherfordium? _____

5. What are two elements in the same column as sodium? _____

6. What are two elements with eight electrons in their outer layer? _____

    _____

7. How many electrons are in the outer layer of nitrogen? _____

8. How many layers of electrons does barium have? _____

9. Name one transition element. _____

10. Would silicon be more likely to react the same way as carbon or chlorine? _____

6 Metals

Silver and gold have I none . . .

🧪 Supply list – Conducting electricity

☐ Flashlight

☐ Battery

☐ Electrical tape (or duct tape)

☐ Copper wire

🧠 What did we learn?

1. What are the six characteristics of most metals?

   a.

   b.

   c.

   d.

   e.

   f.

2. How many valence electrons do most metals have?

3. What is a metalloid?

Taking it further

1. What are the most likely elements to be used in making computer chips?

2. Is arsenic likely to be used as electrical wire in a house?

7 Nonmetals

The rest of the elements

🧪 Supply list – Protecting your teeth

☐ Uncooked egg

☐ Vinegar

☐ Fluoride toothpaste

☐ Nail polish or permanent marker

☐ 2 cups

🏅 Incandescent light bulbs

Optional: How many types of light bulbs do you have in your home? What are they?

🧠 What did we learn?

1. What are some common characteristics of nonmetals?

2. What is the most common state, solid, liquid, or gas, for nonmetal elements?

3. Why are halogens very reactive?

4. Why are noble gases very non-reactive?

🚀 Taking it further

1. Hydrogen often acts like a halogen. How might it act differently from a halogen?

2. Why are balloons filled with helium instead of hydrogen?

| God's Design: Chemistry & Ecology | Properties of Atoms & Molecules | Day 134 | Unit 2 Lesson 8 | Name |

8 Hydrogen

Very reactive

🧪 Supply list – Hydrogenation

☐ Vegetable oil

☐ Margarine

☐ Peanut butter

☐ Prepackaged food labels (ex. crackers, cookies, dry soup, other prepackaged meals)

🏅 Hydrogen fuel cells

Create a poster about hydrogen fuel cells and their benefits.

🧠 What did we learn?

1. What is the most common element in the universe?

2. What is the atomic structure of hydrogen?

3. What is the atomic number for hydrogen?

4. Why is hydrogen sometimes grouped with the alkali metals?

5. Why is hydrogen sometimes grouped with the halogens?

Taking it further

1. Why is hydrogen one of the most reactive elements?

2. Margarine contains only partially hydrogenated oil. What do you suppose fully hydrogenated oils are like?

| God's Design: Chemistry & Ecology | Properties of Atoms & Molecules | Day 136 | Unit 2 Lesson 9 | Name |

9 Carbon

Graphite and diamonds

Supply list – Drawing the carbon cycle; Examining carbon

☐ Drawing paper
☐ Colored pencils
☐ Ceramic plate
☐ Candle
☐ Knife (to scrape carbon from the bottom of the plate)
☐ Matches (to light the candle)

What did we learn?

1. What is the atomic number and atomic structure of carbon?

2. What makes a compound an organic compound?

3. Name two common forms of carbon.
   a.

   b.

4. What is one by-product of burning coal?

🚀 Taking it further

1. How does the carbon cycle demonstrate God's care for His creation?

2. What is the most likely event that caused coal formation?

3. What would happen if bacteria and fungi did not convert carbon into carbon dioxide gas?

| God's Design: Chemistry & Ecology | Properties of Atoms & Molecules | Day 137 | Unit 2 Lesson 10 | Name |

10 Oxygen

A very essential element

🧪 Supply list – Oxygen — needed for burning

☐ Small candle

☐ Dry ice (small piece)

☐ Matches (to light the candle)

☐ Gloves

☐ Glass cup

🎖 Supplies for Challenge – Oxidation

☐ Steel wool

☐ 2 test tubes

☐ Dish

☐ Dish soap

☐ Water

☐ Pencil

🎖 Oxidation

1. What do you observe?

2. Which sample of wool has the most rust?

3. Is the water higher in one tube than in the other tube? Why?

Lesson 10 Properties of Atoms & Molecules // 265

What did we learn?

1. What is the atomic structure of oxygen?

2. How is ozone different from the oxygen we breathe?

Taking it further

1. Why does the existence of ozone in the upper atmosphere show God's provision for life on earth?

2. How do animals in the ocean get the needed oxygen to "burn" the food they eat?

3. Why are oxygen atoms nearly always combined with other atoms?

11 Ionic Bonding

Giving up electrons

Supply list – Atomic models
☐ Colored mini-marshmallows
☐ Toothpicks
☐ Glue

Supplies for Challenge – Ions
☐ Copy of "Name That Ion" worksheet

What did we learn?

1. What is the main feature in an atom that determines how it will bond with other atoms?

2. What kind of bond is formed when one atom gives up electrons and the other atom takes the electrons from it?

3. What is electronegativity?

4. Why are compounds that are formed when one element takes electrons from another called ionic compounds?

5. What are some common characteristics of ionic compounds?

6. Which element has a higher electronegativity, chlorine or potassium?

Taking it further

1. Which column of elements are the atoms in column IA most likely to form ionic bonds with?

2. Use the periodic table of the elements to determine the number of electrons that barium would give up in an ionic bond.

Name _____ Date _____

🎖 Name That Ion Worksheet

Use the periodic table of the elements to help you figure out the names of the following ionic compounds.

NaF

KCl

$CaCl_2$

LiBr

CaS

12 Covalent Bonding

Sharing electrons

🧪 Supply list – More atomic models

☐ Colored mini-marshmallows

☐ Toothpicks

☐ Glue

🏅 Supplies for Challenge – Ionic vs. covalent

☐ Distilled water

☐ 9-volt battery

☐ Copper wire

☐ Sugar

☐ Salt

☐ Baking soda

☐ Olive oil

☐ 4 paper cups

☐ Copy of "Bonding Experiment" Worksheet

🧠 What did we learn?

1. What is a covalent bond?

2. What are some common characteristics of covalent compounds?

3. What is the most common covalent compound on earth?

Taking it further

1. Why do diatomic molecules form covalent bonds instead of ionic bonds?

2. Would you expect more compounds to form ionic bonds or covalent bonds?

Name _____   Date _____

🏅 Bonding Experiment Worksheet

Use a periodic table to determine if each of the following compounds is composed of metals, nonmetals, or both.

Water – H_2O is composed of _____

Baking soda – $NaHCO_3$ is composed of _____

Sugar (sucrose) – $C_{12}H_{22}O_{11}$ is composed of _____

Salt – $NaCl$ is composed of _____

Olive oil (oleic acid) – $C_{18}H_{34}O_2$ is composed of _____

State your hypotheses below and then perform the experiment described in the student manual.

Compound tested	Will it conduct electricity? (Hypothesis)	Did it conduct electricity? (Observations)	Ionic or covalent? (Conclusions)
Distilled water			
Baking soda			
Sugar			
Salt			
Olive oil			

Were any of your guesses wrong? If so, try to find out about that substance and see if it is really ionic or covalent and try to understand why each substance acted the way it did. Checking your hypothesis and understanding your results is an important part of scientific experimentation.

Lesson 12 **Properties of Atoms & Molecules** // 273

13 Metallic Bonding

Sharing on a large scale

🧪 Supply list – Metal models
☐ Colored mini-marshmallows
☐ Toothpicks
☐ Glue

🏅 Supplies for Challenge – Bonding characteristics
☐ Copy of "Bonding Characteristics" Worksheet

🧠 What did we learn?
1. What is the free electron model?

2. How many valence electrons do metals usually have?

3. What are common characteristics of metallic compounds?

🚀 Taking it further
1. Why don't metals form ionic or covalent bonds?

2. Would you expect semiconductors to form metallic bonds?

Name _____ Date _____

Bonding Characteristics Worksheet

Place the following characteristics in the appropriate column(s) below. Some characteristics belong to more than one type of bonding.

A. Electrons are transferred.

B. Free electrons.

C. Electrons are shared.

D. Forms between metals.

E. Forms between nonmetals.

F. Forms between metals and nonmetals.

G. Forms between elements with different electronegativities.

H. Forms between elements with similar high electronegativities.

I. Forms between elements with similar low electronegativities.

J. Solids with high melting points.

K. Solids with low melting points, liquids, or gases.

L. Conducts electricity.

M. Does not conduct electricity.

N. Does not easily dissolve.

O. Easily dissolves in water.

Ionic bonding	Covalent bonding	Metallic bonding

14 Mining & Metal Alloys

Making it stronger

🧪 Supply list – Polishing silver
☐ Tarnished silver object
☐ Silver polish
☐ Soft cloth

🎖 Supplies for Challenge – Alloys
☐ Copy of "Common Alloys" Worksheet

🧠 What did we learn?

1. What element is combined with most metals to form metal ore?

2. What must be done to metal oxides to obtain pure metal?

3. What is an alloy?

4. Why are alloys produced?

🚀 Taking it further

1. Do you think chromium would be added to steel that is going to be used in saw blades? Why or why not?

2. Is oxidation of metal always a bad thing?

Name _____ Date _____

🎖 Common Alloys Worksheet

For each common alloy below, list the metals that are combined together to form that alloy.

Bronze _____

Brass _____

Steel _____

Solder _____

Duraluminium _____

Magnalium _____

Pewter _____

Sterling silver _____

Stainless steel _____

15 Crystals

Sparkling like diamonds

🧪 Supply list – Growing crystals; Opening a geode (optional)

☐ Table salt

☐ 2 plates (1 for table salt experiment, 1 for Epsom salt experiment)

☐ Epsom salt

☐ Scissors

☐ Dark construction paper

☐ Small pan

☐ Water

☐ Access to stove

☐ Optional activity: geode

🎖 Supplies for Challenge – Hydrates

☐ Plaster of Paris

☐ Modeling clay

☐ Water

☐ A pet (optional)

🧠 What did we learn?

1. What is a crystal?

2. How do crystals form?

3. What is an artificial gem?

4. Where would you look to find crystals?

🚀 Taking it further

1. Why are naturally occurring gems more valuable than artificial gems when many are made from the same materials?

2. Why is a saturated solution better for forming crystals?

3. What are some ways you use crystals in your home?

| God's Design: Chemistry & Ecology | Properties of Atoms & Molecules | Day 147 | Unit 3 Lesson 16 | Name |

16 Ceramics

Making it with clay

🧪 Supply list – Fun with clay

☐ Polymer clay (Fimo®, Sculpey®, etc.)

🏅 Bioceramics

Tell your teacher about bioceramics. Be sure to note any beneficial uses.

🧠 What did we learn?

1. What is ceramic?

2. What are some examples of traditional ceramics?

3. What makes ceramics hard?

4. What are some advantages of modern ceramics?

Taking it further

1. Why are the tiles on the space shuttle made of ceramic?

2. Why are crystalline structures stronger than noncrystalline structures?

17 Chemical Reactions

Changing from one thing to another

🧪 Supply list – Fire extinguisher in a jar

☐ Birthday candle

☐ Vinegar

☐ Modeling clay

☐ Baking soda

☐ Jar

☐ Matches (to light the candle)

🏅 Supplies for Challenge – Temperature & surface area

☐ 6 clear cups   ☐ Pan

☐ 6 Alka-Seltzer® tablets   ☐ Access to stove

☐ Water   ☐ Paper

☐ Ice   ☐ Spoon

☐ Stopwatch   ☐ Copy of "Reaction Rate Experiment" Worksheet

🧠 What did we learn?

1. What is a chemical reaction?

2. What are the initial ingredients in a chemical reaction called?

3. What are the resulting substances of a chemical reaction called?

🚀 Taking it further

1. How might you speed up a chemical reaction?

2. A fire hose usually sprays water on a fire to put it out. Water does not deprive the fire of oxygen, so why does water put out a fire?

3. What chemical reaction do you think is taking place in the making of a loaf of bread?

Name _____   Date_____

🏅 Reaction Rate Experiment Worksheet

State your hypotheses below and then perform the experiment as described in your student manual.

Hypotheses

1. Higher temperature will _____ the reaction rate.

2. Lower temperature will _____ the reaction rate.

3. Greater surface area will _____ the reaction rate.

Experiment 1: Effects of Temperature

	Hot	Room temperature	Cold
Time to dissolve			
Observations			

Which tablet dissolved fastest? _____

Which tablet dissolved slowest? _____

Experiment 2: Effects of Surface Area

	Whole tablet	Broken tablet	Crushed tablet
Time to dissolve			
Observations			

Which tablet dissolved fastest? _____

Which tablet dissolved slowest? _____

Lesson 17 **Properties of Atoms & Molecules**

18 Chemical Equations

Describing how it works

🧪 Supply list – Chemical equations
☐ Copy of "Understanding Chemical Equations" Worksheet

🏅 Supplies for Challenge – Reactants and products
☐ Copy of "Reactants and Products" Worksheet

🧠 What did we learn?

1. What is a chemical equation?

2. What are the elements or compounds on the left side of a chemical equation called?

3. What are the elements or compounds on the right side of a chemical equation called?

🚀 Taking it further

1. Why is it helpful to use chemical equations?

Name _____ Date _____

🧪 Understanding Chemical Equations Worksheet

For each chemical reaction that is shown, follow the steps below to help you write the correct chemical equation for each reaction.

Step 1: Write the atomic symbol for each element below the picture of the atom or molecule.

Step 2: Count how many of each type of atom is in each picture. If there is more than one atom bonded together (touching), write that number after the atomic symbol as a subscript (smaller and lower down than the letters). If there are more than one of the same atom not bonded together, or more than one of the same molecule, write that number in front of the symbol for that atom or molecule.

Step 3: Draw an arrow (⟶) between the reactants and the products.

Step 4: Draw a plus sign (+) between the atoms or molecules on each side of the arrow.

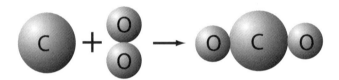

_____

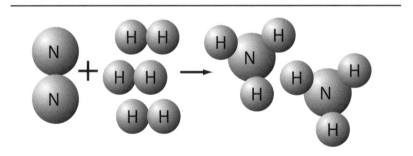

_____

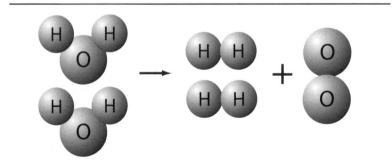

_____

288 // God's Design: Chemistry & Ecology

Name _____   Date_____

🎖 Reactants & Products Worksheet

A chemical equation is a way to describe what is happening in a chemical reaction. Remember that the kind and number of elements on one side of the arrow (reactants) has to be the same as the elements on the other side of the arrow (products).

Match the reactants with the correct products.

1. $4\ Al + 3\ O_2 \longrightarrow$ A. $Li_2SO_4 + 2\ H_2O$

2. $H_2SO_4 + 2\ LiOH \longrightarrow$ B. $2\ Al_2O_3$

3. $4\ NH_3 + 3\ O_2 \longrightarrow$ C. $2\ N_2 + 6\ H_2O$

4. $P_4 + 10\ Cl_2 \longrightarrow$ D. $C + O_2$

5. $CO_2 \longrightarrow$ E. $4\ PCl_5$

6. $H + OH \longrightarrow$ F. H_2O

7. $2\ KClO_3 \longrightarrow$ G. $H_2 + 2\ NaOH$

8. $2\ Na + 2\ H_2O \longrightarrow$ H. $2\ KCl + 3\ O_2$

Which of the above equations represent decomposition reactions? _____

Which of the above equations represent composition reactions? _____

Which of the above equations represent single displacement reactions? _____

Which of the above equations represent double displacement reactions? _____

Lesson 18 **Properties of Atoms & Molecules**

| God's Design: Chemistry & Ecology | Properties of Atoms & Molecules | Day 152 | Unit 4 Lesson 19 | Name |

19 Catalysts

Speeding things up

Supply list – Catalysts & inhibitors

☐ Potato
☐ Hydrogen peroxide
☐ Apple
☐ Lemon juice
☐ Drinking glass
☐ Knife
☐ Brush

Catalysts & inhibitors

1. What do you think will happen to the slices with the lemon juice?

2. What do you think will happen to the slices without the lemon juice?

3. What differences do you see between the slices with the lemon juice and those without?

Types of catalysts

Draw a chart listing out the different types of catalysts you learned about today.

What did we learn?

1. What is a catalyst?

2. How does a catalyst work?

3. What is an inhibitor?

4. What is an enzyme?

Taking it further

1. Why is it important that living cells have enzymes?

2. Are catalysts always good?

| God's Design: Chemistry & Ecology | Properties of Atoms & Molecules | Day 153 | Unit 4 Lesson 20 | Name |

20 Endothermic & Exothermic Reactions

What happens to the heat?

🧪 Supply list – Endothermic & exothermic reactions

- ☐ 5 uncooked eggs
- ☐ Small sauce pan
- ☐ Thermometer
- ☐ Timer or clock
- ☐ Steel wool (no soap)
- ☐ Vinegar (room temperature)
- ☐ Jar with lid (thermometer must fit inside the jar with the lid on)
- ☐ 5 pieces of paper
- ☐ Water
- ☐ Access to a stove

🏅 Supplies for Challenge – Energy in a reaction

- ☐ Copy of "Endothermic or Exothermic?" Worksheet
- ☐ Alka-Seltzer® tablets
- ☐ Thermometer
- ☐ Styrofoam™ cup
- ☐ Water
- ☐ Stopwatch

🧠 What did we learn?

1. What is an exothermic reaction?

2. What is an endothermic reaction?

🚀 Taking it further

1. If a chemical reaction produces a spark, is it likely to be an endothermic or exothermic reaction?

2. How do photosynthesis and digestion reveal God's plan for life?

3. If the temperature of the product is lower than the temperature of the reactants, was the reaction endothermic or exothermic?

Name _____   Date _____

🎖 Endothermic or Exothermic? Worksheet

You will be combining Alka-Seltzer® with water. Do you expect this reaction to be endothermic or exothermic?
_____

Heat a cup of water to boiling. Let the water cool for about two minutes. Pour the water into a Styrofoam™ cup. Measure the temperature of the water and record it on the chart below. Add two Alka-Seltzer® tablets to the water, and then record the temperature every 10 seconds for one minute.

Time	Temperature
Initial	
10 seconds	
20 seconds	
30 seconds	
40 seconds	
50 seconds	
60 seconds	

Did the temperature of the water increase or decrease during the reaction? _____

Was the reaction endothermic or exothermic? _____

Graph your data below to visually show what happened during the reaction.

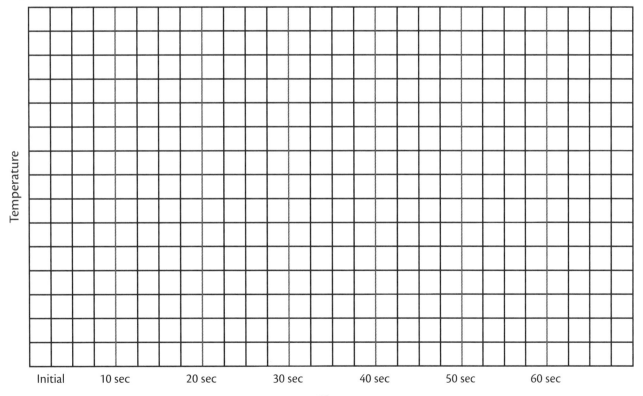

Lesson 20 **Properties of Atoms & Molecules** // 295

| God's Design: Chemistry & Ecology | Properties of Atoms & Molecules | Day 155 | Unit 5 Lesson 21 | Name |

21 Chemical Analysis

What is it made of?

🧪 Supply list – Making an acid/base indicator
☐ Purple cabbage

☐ Microwave oven or sauce pan and stove

☐ Water

🏅 Supplies for Challenge – Chemical analysis methods
☐ Research materials on chemical analysis

🧠 What did we learn?

1. What is chemical analysis?

2. List three different types of chemical analysis.

    a.

    b.

    c.

3. What is a chemical indicator?

4. What is the pH scale?

5. What does a pH of 7 tell you about a substance?

Taking it further

1. Why is it important to periodically test the pH of swimming pool water?

2. Name at least one other use for testing pH of a liquid.

22 Acids

Does it burn?

🧪 Supply list – Testing for acids

☐ Lemon juice

☐ Vinegar

☐ Clear soda (ex. lemon lime)

☐ Milk

☐ Saliva

☐ Cabbage indicator from lesson 21

🧪 Testing for acids

1. What color did the indicator become when mixed with each of these items?

2. Which items are acidic?

🏅 Supplies for Challenge – Displacement reaction

☐ Jar with lid

☐ 15 pennies

☐ 1 steel paper clip

☐ Salt

☐ Vinegar

🏅 Displacement reaction

After 15 minutes:

1. How do they look?

2. What do you observe happening in the jar?

After 30 minutes:

1. How do they look?

2. What do you observe happening in the jar?

After 60 minutes:

1. How do they look?

2. What do you observe happening in the jar?

What did we learn?

1. What defines a substance as an acid?

2. What is a hydronium ion?

3. How is a weak acid different from a strong acid?

4. What are some common characteristics of an acid?

5. How can you tell if a substance is an acid?

Taking it further

1. Why is saliva slightly acidic?

2. Would you expect water taken from a puddle on the forest floor to be acidic, neutral, or basic? Why?

3. What would you expect to be a key ingredient in sour candy?

23 Bases

The opposite of acids

Supply list – Testing for bases

☐ Ammonia (clear)
☐ Soap
☐ Anti-acid (tablets or liquid)
☐ Baking soda
☐ Toothpaste
☐ Cabbage indicator from lesson 21

Supplies for Challenge – Acid/base titration

☐ Ammonia
☐ Eyedropper
☐ Distilled water
☐ Measuring cup
☐ Measuring spoon
☐ Clear glass
☐ Vinegar
☐ Acid/base indicator

What did we learn?

1. What defines a substance as a base?

2. What is a hydroxide ion?

3. How is a weak base different from a strong base?

4. What are some common characteristics of a base?

5. How can you tell if a substance is a base?

🚀 Taking it further

1. If you spill a base, what should you do before trying to clean it up?

2. Do you think that strontium (Sr) is likely to form a strong base? Why or why not?

24 Salts

Pass the salt, please.

🧪 Supply list – Acid + base = salt + water

☐ Water

☐ Measuring cups and spoons

☐ Vinegar

☐ Baking soda

☐ Cup

☐ Cotton swabs

🎖 Supplies for Challenge – Acid/base reactions

☐ Copy of "Acid/Base Reactions" Worksheet

🧠 What did we learn?

1. How is a salt formed?

2. What are two common characteristics of salts?
    a.

    b.

3. How are salt families named?

4. Name three salt families.

   a.

   b.

   c.

🚀 Taking it further

1. What do you expect to be the results of combining vinegar and lye?

2. Why are some salts still acidic or basic?

Name _____   Date_____

🎖 Acid/Base Reactions Worksheet

Below are the chemical equations for several acid/base reactions. For each reaction, identify which substance is the acid, which is the base, and which is the salt. It will help if you remember that acids give up a hydrogen ion (H^+) when they combine with bases, and bases give up a hydroxide ion (OH^-).

1. $HClO_3 + KOH \longrightarrow KClO_3 + H_2O$

    The acid is _____   The salt is _____

    The base is _____

2. $HBr + Ca(OH)_2 \longrightarrow CaBr_2 + H_2O$

    The acid is _____   The salt is _____

    The base is _____

Recall that an alternate definition for an acid is a *proton donor* and an alternate definition for a base is a *proton acceptor*. Identify the acid and base in the following equations based on which substance is losing a hydrogen atom and which is gaining one.

3. $H_2SO_4 + 2NH_3 \longrightarrow 2NH_4^+ + SO_4^{2-}$

    The acid is _____

    The base is _____

4. $HI + H_2O \longrightarrow H_3O^+ + I^-$

    The acid is _____

    The base is _____

25 Biochemistry

The chemistry of life

Supply list – A balanced diet
☐ Whatever food you have in your kitchen (specifically with food labels)

Supplies for Challenge – Enzymes
☐ Box of gelatin mix
☐ Fresh pineapple juice (do not use canned or frozen)
☐ Vinegar
☐ 4 cups
☐ Measuring spoon
☐ Sauce pan
☐ Access to a refrigerator
☐ Access to a stove
☐ Copy of "Enzyme Reaction" Worksheet

What did we learn?

1. List at least two chemical functions performed inside living creatures.
   a.

   b.

2. What is the chemical reaction that takes place during photosynthesis?

3. What is the main chemical reaction that takes place during digestion?

4. What substance is necessary for nearly every chemical reaction in living things?

5. Name the three major chemicals your body needs that are found in the foods we eat.

   a.

   b.

   c.

🚀 Taking it further

1. Why did God design your body to have enzymes?

2. With what you know about chemical processes, why do you think it is important to brush your teeth after you eat?

3. Can you think of other chemical processes in your body besides the ones mentioned in this lesson?

Name _____   Date _____

🎖 Enzyme Reaction Worksheet

Write what you think will happen in each cup of gelatin in the hypothesis boxes below. Then observe each cup of gelatin after the given time and write your observations in the chart. After 3 hours, answer the questions at the bottom of the page.

	Cup 1 (Protease)	Cup 2 (Protease + acid)	Cup 3 (Heated protease)	Cup 4 (No protease)
Hypothesis				
Observations after 30 minutes				
Observations after 60 minutes				
Observations after 90 minutes				
Observations after 2 hours				
Observations after 3 hours				

Conclusions

1. What effects did the plain pineapple juice have on the gelatin?

   _____

2. What effects did changing the pH have on the way the juice affected the gelatin?

   _____

3. What effects did heating the juice have on the way the juice affected the gelatin?

   _____

4. Why was cup 4 necessary?

   _____

Lesson 25 **Properties of Atoms & Molecules**

| | *God's Design: Chemistry & Ecology* | Properties of Atoms & Molecules | Day 163 | Unit 6 Lesson 26 | Name |

26 Decomposers

Ultimate recycling

🧪 Supply list – The nitrogen cycle

☐ Drawing paper

☐ Colored pencils

Example:

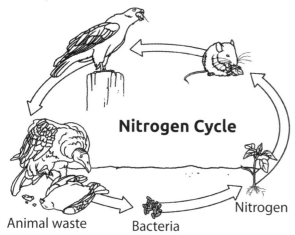

🏅 Supplies for Challenge – Rate of decomposition

☐ Banana

☐ 3 plastic zipper bags

☐ Baking yeast

☐ Marker

☐ Copy of "Rate of Decomposition" worksheet

🧠 What did we learn?

1. What is a scavenger?

2. What is a decomposer?

Lesson 26 **Properties of Atoms & Molecules** // 311

3. What is this way of recycling nitrogen called?

🚀 Taking it further

1. Why are decomposers necessary?

2. Were there animal scavengers in God's perfect creation, before the Fall of man?

3. Explain how a compost pile allows you to participate in the nitrogen cycle.

Name _____ Date _____

🏅 Rate of Decomposition Worksheet

State your hypothesis below and then perform the experiment described in your student manual.

Hypothesis

Which conditions do you think will produce the quickest decomposition?

_____

	Cold and dark	Warm and dark	Warm and light
Initial condition of banana			
Day 1			
Day 2			
Day 3			
Day 4			
Day 5			
Day 6			
Day 7			

Conclusion

1. Which conditions actually produced the quickest decomposition?

_____

2. Why are many foods kept in the refrigerator or freezer?

_____

27 Chemicals in Farming

Helping plants grow

🧪 Supply list – The effects of fertilizer

☐ 2 identical plants—a fast growing plant like mint is a good choice

☐ Plant food

🧪 The effects of fertilizer

1. Based on what you have learned, which plant would you expect to grow faster? Why?

🎖 Organic farming

Research organic farming. You may even want to compare similar food items — one organic and non-organic. Based on your research, answer the following questions:

1. Controlling pests—are organic methods effective?
2. Productivity—which way produces more crops?
3. Labor required—which way requires more labor?
4. Genetically modified organisms—are they good or bad?
5. The environment—which farming method is friendlier to the environment?
6. Food quality—which is better?
7. Food health—which is healthier?

🧠 What did we learn?

1. What are three ways that farmers ensure their soil will have enough nutrients for their crops?

    a.

    b.

    c.

2. What is hydroponics?

3. How are chemicals used in farming other than for nutrients for the plants?

4. How is an organic farm different from other farms?

🚀 Taking it further

1. Why did the farmers let cattle graze on their land once every fourth year in the Norfolk 4-course plant rotation method?

2. How does hydroponics replace the role of soil in plant growth?

28 Medicines

Chemical compounds that affect your body

🧪 Supply list – Common herbs

☐ Bread

☐ Butter or margarine

☐ Garlic powder

☐ Ginger ale

🧠 What did we learn?

1. Why are chemicals used as medicines?

2. What were the earliest recorded medicines?

3. What was Sir Alexander Fleming's important discovery?

🚀 Taking it further

1. If plants have the potential of supplying new medicines, where might a person look to find different plants?

2. What other sources might there be for discovering new medicines?

29 Perfumes

What's that smell?

🧪 Supply list – Making your own perfume

☐ Jar with lid

☐ Rubbing alcohol

☐ 15 whole cloves

🏅 Supplies for Challenge – Scents (several of the following)

☐ Ginger root

☐ Mint leaves

☐ Peppermint oil

☐ Cinnamon sticks

☐ Dried fruit

☐ Flower petals

☐ Allspice

☐ Almond extract

☐ Vanilla extract

🧠 What did we learn?

1. What is a perfume?

2. What must be removed from flower petals to make perfume?

3. Describe the two main methods for removing oil from flower petals.

   a.

   b.

🚀 Taking it further

1. Why should you test a new perfume on your skin before you buy it?

2. Why wasn't it necessary to use one of the methods described in the lesson to make your homemade perfume?

30 Rubber

Do you have a rubber band?

🧪 Supply list – Playing with rubber

☐ Latex balloon

☐ Wide rubber band

☐ Marker

🏅 Polymers

Discuss with your teacher what polymers are and examples of some animals that produce it. What are some benefits you see in it?

🧠 What did we learn?

1. What is natural rubber made from?

2. What is synthetic rubber made from?

3. What is vulcanization?

4. What is a polymer?

🚀 Taking it further

1. Why is it difficult to recycle automobile tires?

2. What advantages and disadvantages are there to using synthetic rubber instead of natural rubber?

| God's Design: Chemistry & Ecology | Properties of Atoms & Molecules | Day 172 | Unit 7 Lesson 31 | Name |

31 Plastics

The wonder material

Supply list – Chemical word search
☐ Copy of "Chemical Word Search"

Supplies for Challenge – Polymer ball
☐ Borax powder
☐ White glue
☐ Cornstarch
☐ Markers
☐ Water
☐ Plastic zipper bag
☐ 2 cups

What did we learn?

1. What is plastic?

2. What was celluloid, the first artificial polymer, made from?

3. What is the difference between thermoplastic and thermosetting resin?

4. Why are people concerned about throwing plastic items away?

5. What does the recycling number on a plastic item mean?

Lesson 31 **Properties of Atoms & Molecules** // 323

6. Why are plastic bags usually recycled separately from other plastics?

🚀 Taking it further

1. Name three ways that plastic is used in sports.

    a.

    b.

    c.

2. What advantages do plastic items have over natural materials?

Chemical Word Search

Find the following words in the puzzle below. Words may be horizontal, vertical, or diagonal, including backward.

Acid	Bond	Fat	Photosynthesis	Rubber
Alloy	Carbohydrate	Indicator	Plastic	Salt
Atom	Digestion	Decomposer	Polymer	Synthetic
Base	Distillation	Perfume	Protein	Vulcanization

S	A	L	T	V	U	K	A	T	M	A	T	O	M	E
C	S	T	R	I	S	Y	N	T	H	E	T	I	C	I
C	P	E	R	F	U	M	E	O	R	E	A	D	F	U
A	E	S	S	A	K	T	V	I	X	Y	J	I	D	Y
R	P	H	O	T	O	S	Y	N	T	H	E	S	I	S
B	L	R	P	S	B	O	N	D	I	U	W	T	G	Z
O	A	G	O	W	T	H	E	I	S	C	N	I	E	M
H	S	U	L	T	B	N	M	C	D	F	H	L	S	I
Y	T	V	Y	N	E	W	B	A	S	E	B	L	T	Z
D	I	S	M	Q	U	I	I	T	A	T	T	A	I	S
R	C	R	E	J	E	P	N	O	B	L	I	T	O	M
A	R	K	R	E	B	B	U	R	C	A	L	I	N	Y
T	V	U	L	C	A	N	I	Z	A	T	I	O	N	S
E	A	A	C	I	D	D	H	G	Y	T	R	N	Y	O
W	P	R	U	F	R	E	S	O	P	M	O	C	E	D

Lesson 31 **Properties of Atoms & Molecules** 325

| God's Design: Chemistry & Ecology | Properties of Atoms & Molecules | Day 173 | Unit 7 Lesson 32 | Name |

32 Fireworks

Is it the Fourth of July?

🧪 Supply list – Designing a fireworks display
☐ Construction paper
☐ Various colors of glitter
☐ Glue

🎖 Supplies for Challenge – Colored flames
☐ Several pinecones
☐ Table salt
☐ Epsom salts (found in the medicine section of the store)
☐ Potassium chloride (used as a salt substitute — may be found in the spice section of grocery store)
☐ Borax and calcium chloride (may be found with laundry/cleaning supplies)
☐ Copper sulfate (found where swimming pool supplies are sold)
☐ Several containers in which to soak the pinecones
☐ Sodium chloride
☐ Copper chloride
☐ Magnesium sulfate

🧠 What did we learn?

1. What are the key ingredients in a fireworks shell?

2. Why does a fireworks shell have two different black powder charges?

3. How do fireworks generate flashes of light?

Lesson 32 **Properties of Atoms & Molecules** ✦✦ 327

4. What determines the color of the firework?

🚀 Taking it further

1. How can a firework explode with one color and then change to a different color?

2. Why would employees at a fireworks plant have to wear only cotton clothing?

33 Rocket Fuel

Do you need a rocket scientist?

🧪 Supply list – Balloon rocket

☐ Balloon

☐ String

☐ 2 chairs

☐ Drinking straw

☐ Tape

🧠 What did we learn?

1. What is combustion?

2. What two elements are combined in most modern rocket fuel?

    a.

    b.

3. What compound is produced in this reaction?

4. How does combining oxygen and hydrogen produce lift?

5. What is Newton's third law of motion?

 Taking it further

1. Why is oxygen and hydrogen a better choice for rocket fuel than kerosene was?

| God's Design: Chemistry & Ecology | Properties of Atoms & Molecules | Day 176 | Unit 7 Lesson 34 | Name |

34 Fun With Chemistry: Final Project

Understanding chemical reactions

🧪 Final Project Supply list

☐ Milk (not skim)

☐ Paper towel

☐ Food coloring

☐ Dish

☐ Sink

☐ Water

☐ Large glass

☐ Water soluble markers

☐ Liquid dish soap

☐ White glue

☐ Unused disposable baby diaper

☐ Liquid starch

☐ Scissors

☐ Plastic zipper bag

☐ Eyedropper

☐ Copy of "Fun With Chemistry" Worksheet

🏅 More experiments

Note: This challenge is optional.

🧠 What did we learn?

1. What was your favorite chemical reaction?

Lesson 34 **Properties of Atoms & Molecules** // 331

2. Why did you like that reaction?

🚀 Taking it further

1. What do you think will happen if you use skim milk in the first activity?

2. What colors would you expect to see separate out of orange ink? Brown ink?

3. Why is it important not to inhale the sodium polyacrylate from the diaper?

Name _____   Date _____

🧪 Fun With Chemistry Worksheet

Activity 1: Moving Molecules

Beginning ingredients	What I did	What I observed

This is why I think the colors swirled together after I added the soap: _____
_____

Activity 2: More Moving Molecules

Colors observed in the black ink	Colors observed in the green ink	Colors observed in the _____ ink

This is why I think the colors separated as they moved up the paper: _____
_____

Activity 3: Super Absorbent Molecules

Observations of powder before adding water	Observations of powder after adding water

This is why I think the powder changed its appearance: _____
_____

Activity 4: Making Your Own Goop

Observations of ingredients before mixing	Observations of ingredients after mixing

This is what I think happened in the chemical reaction: _____
_____

Lesson 34 **Properties of Atoms & Molecules**

| God's Design: Chemistry & Ecology | Properties of Atoms & Molecules | Day 178 | Unit 7 Lesson 35 | Name |

35 Conclusion

Appreciating our orderly universe

 **Supply list**

☐ Bible

Matter Quizzes and Final Exam

for Use with

Properties of Matter

(*God's Design: Chemistry & Ecology*)

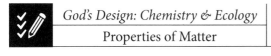 God's Design: Chemistry & Ecology — Properties of Matter | Quiz 1 | Scope: Lessons 1–4 | Total score: ____ of 100 | Name

Experimental Science

Number the steps of the scientific method in the correct order (5 points each).

A. _____ Ask a question.

B. _____ Learn about something/Make observations.

C. _____ Share your results.

D. _____ Design a test and perform it.

E. _____ Make a hypothesis.

F. _____ Check your results/Is your hypothesis right?

Mark each statement as either True or False (6 points each).

1. _____ You must always have a correct hypothesis.

2. _____ It is important to control variables in your experiments.

3. _____ Qualitative observations always use numbers.

4. _____ Quantitative observations can be more useful to scientists than qualitative observations.

5. _____ It is usually easier to make conversions between units in the metric system than in the Old English/American system.

6. _____ A millimeter is smaller than a meter.

7. _____ A graduated cylinder should be used to measure mass.

8. _____ God has established laws to govern how chemicals react with each other.

9. _____ Science can always tell us why things happen.

10. _____ Matter has no mass.

Short answer (10 points):

11. Describe what chemistry is the study of.

_____

_____

_____

Challenge questions

Short answer (8 points each):

1. Is the measurement of the intensity of light from a distant star origins science or observation science?

2. Is the use of distant starlight to date the universe an example of origins science or observation science?

3. Why shouldn't you look through the eyepiece while lowering the objective on a microscope?

4. How are microscopes similar to telescopes?

5. How are microscopes different from telescopes?

Match the scale with what phenomenon it describes (10 points each).

6. _____ Mohs scale

7. _____ Fujita scale

8. _____ Saffir-Simpson scale

9. _____ Beaufort scale

10. _____ Richter scale

11. _____ Mercalli scale

A. Tornado intensity

B. Wind

C. Mineral hardness

D. Earthquake intensity

E. Hurricane intensity

F. Damage done by an earthquake

Measuring Matter

Match the term with its definition (6 points each).

1. _____ The amount of a substance.
2. _____ How strongly something is pulled on by gravity.
3. _____ Matter cannot be created or destroyed.
4. _____ How much space matter occupies.
5. _____ How much mass is in a particular volume.
6. _____ The ability for one substance to float in another.
7. _____ Used to measure mass.
8. _____ Used to measure weight.
9. _____ A material that is denser than lead.
10. _____ Only common material to become less dense when frozen.

A. Conservation of mass
B. Density
C. Mass
D. Volume
E. Spring scale
F. Balance
G. Buoyancy
H. Weight
I. Water
J. Gold

Short answer (8 points each):

11. Explain how the water you drink today could be the same water a dinosaur drank thousands of years ago. _____

12. Explain what happens to nitrogen in the soil and in plants that demonstrates conservation of mass. _____

13. If an object floats in one liquid but sinks in another, what does that tell you about the densities of the two liquids?

14. How would you determine the volume of a toy car? _____

15. How are buoyancy and density related? _____

Challenge questions

Short answer (20 points each):

1. What are two units for measuring mass?

2. What are two units for measuring weight?

3. If you perform an experiment, and the mass of the resulting substance is less than the mass of what you started with, what is one likely explanation?

4. Which is likely to be more dense, a one-inch cube of steel or a one-inch cube of wood?

5. If you are traveling in a car with a helium balloon and the driver suddenly puts on the brakes, what will happen to your body, and what will happen to the balloon?

States of Matter

Fill in the blank with the correct term from below (some are used more than once) (5 points per blank).

Adding heat   Solid   Removing heat   Liquid   Gas

1. The three states of matter are _____, _____, and _____.
2. _____ causes the molecules in matter to move more quickly.
3. _____ causes the molecules in matter to move more slowly.
4. _____ is required to change a gas into a liquid.
5. _____ is required to change a solid into a liquid.

Write S beside the statement if it describes a property of a solid, L if it describes a liquid, and G if it describes a gas. Some statements describe more than one state of matter (3 points each).

6. _____ Molecules are close together.
7. _____ Molecules are far apart.
8. _____ It takes on the shape of its container.
9. _____ Molecules move very quickly.
10. _____ Molecules slide over each other.
11. _____ Easily compressed.
12. _____ Has a defined shape.
13. _____ Has a defined volume.
14. _____ Molecules only vibrate.
15. _____ Not easily compressed.

Mark each statement as either True or False (5 points each).

16. _____ Thick liquids have a high viscosity.
17. _____ As the temperature of a gas increases, its volume decreases.
18. _____ As the pressure of a gas increases, its volume decreases.
19. _____ A ball will usually bounce better on a warm day than on a cold one.
20. _____ Molecules in a viscous liquid are not strongly attracted to each other.
21. _____ There is a direct relationship between the temperature of a gas and its volume.
22. _____ Crystals are more likely to form when a solid cools slowly.

Challenge questions

Mark each statement as either True or False (10 points each).

1. _____ Solid water is less dense than liquid water.
2. _____ Glass can be classified as an amorphous solid.
3. _____ Evaporation requires that a liquid be heated to the boiling point.
4. _____ Diffusion occurs as molecules move from an area of lower concentration to an area of higher concentration.
5. _____ Glass does not have a definite boiling point.
6. _____ All solids are denser than their liquid form.
7. _____ Evaporation is slower on windy days.
8. _____ Increasing surface area increases evaporation rate.

Identify each of the following changes as a chemical change (C) or a physical change (P) (4 points each).

9. _____ Adding water to orange juice.
10. _____ Shredding a piece of paper.
11. _____ Taking aspirin for a headache.
12. _____ Burning a candle.
13. _____ Shooting off fireworks.

Quiz 4 | Scope: Lessons 16–21 | Total score: ___ of 100 | Name

Classifying Matter

Match the terms below with their correct definition or description (6 points each).

A. Element   D. Homogeneous   G. Pasteurization   J. Foam
B. Compound   E. Heterogeneous   H. Water
C. Mixture   F. Structural Integrity   I. Homogenization

1. _____ A combination of two or more pure substances where each keeps its own properties—a new substance is *not* formed.
2. _____ A liquid with air bubbles trapped in it.
3. _____ A substance made when two or more elements combine chemically.
4. _____ The process of heating a mixture to kill the bacteria in it.
5. _____ A substance that cannot be broken down chemically.
6. _____ A mixture where the substances are thoroughly mixed up.
7. _____ A mixture where the substances are not evenly mixed up.
8. _____ A nearly universal solvent.
9. _____ The process of breaking up fat into tiny pieces that can remain suspended.
10. _____ The ability of fat molecules to keep air molecules suspended.

Short answer (10 points each):

11. Explain why whipped cream eventually melts into a pool of white liquid.

12. Give an example showing that a compound does not act like the elements that it is made from.

13. Explain why water is considered by many to be a nearly universal solvent.

14. What elements are found in the compound CH_4? _____

Challenge questions

Short answer (Each answer worth 10 points each - note that question 2 and 4 have 3 answers each):

1. What is a mineral?
   _____

2. Name three groups of minerals. _____
   _____   _____

3. What is a native mineral?
   _____

4. List three ways to separate substances in a mixture. _____
   _____   _____

5. Briefly explain how milk is turned into cheese.
   _____

6. What instrument is used to separate blood cells from plasma?
   _____

| God's Design: Chemistry & Ecology — Properties of Matter | Quiz 5 | Scope: Lessons 22–28 | Total score: ____of 100 | Name |

Solutions

Mark each statement as either True or False (5 points each).

1. _____ All solutions are mixtures.

2. _____ All mixtures are solutions.

3. _____ A saturated solution can dissolve more solute.

4. _____ True solutions do not settle out.

5. _____ Milk is a true solution.

6. _____ Temperature affects how fast substances dissolve.

7. _____ Surface areas affect how fast substances dissolve.

8. _____ A dilute solution has a high amount of solute.

9. _____ The boiling point of a solution is affected by concentration.

10. _____ Cold liquids can suspend more gas than warmer liquids.

Short answer (10 points each):

11. Describe why decreasing temperature decreases the solubility of a solid in a liquid.

12. Describe why increased pressure increases the solubility of a gas in a liquid.

13. What is a precipitate?

14. Why is salt added to ice when freezing ice cream?

15. What is likely to happen to a car without antifreeze in the radiator?

Challenge questions

Short answer (10 points each):

1. If a substance easily dissolves in water, would you expect it to easily dissolve in oil?

2. Why does soap easily dissolve in both water and oil?

3. Name one commercial application for an emulsion.

4. How does the concentration of salt affect the boiling point of water?

5. Why does salt affect the boiling point this way?

6. How does temperature affect the density of seawater?

7. Why is hard water considered a problem?

8. Explain why cake is considered an emulsion.

9. Why is water often called the universal solvent?

10. Explain why it is easy to float in the Dead Sea.

| God's Design: Chemistry & Ecology — Properties of Matter | Quiz 6 | Scope: Lessons 29–33 | Total score: ____ of 100 | Name |

Food Chemistry

Choose the best answer for each question (10 points each).

1. _____ Which is the most popular drink in the world?
   A. Tea B. Water C. Soft drink D. Coffee

2. _____ What are 70% of all soft drinks sweetened with?
   A. Corn syrup B. Saccharine C. Sugar D. Aspartame

3. _____ To guarantee that each soft drink tastes the same, what must a company do?
   A. Purify the water B. Guard its recipe C. Test the solutions D. All of these

4. _____ Which of the following is not used to make soft drinks?
   A. Corn syrup B. Flavorings C. Carbon dioxide D. Oxygen

5. _____ What accounts for the perceived flavor of a food?
   A. Chemical compounds B. Color C. Texture D. All of these

6. _____ Which civilization is believed to be the first to enjoy chocolate?
   A. Egyptians B. Aztecs C. Romans D. Greeks

7. _____ What is the purpose of fermenting vanilla beans?
   A. Keep harvesters busy B. Change the color C. Develop flavor D. Kill bacteria

8. _____ Which of the following is a natural flavor?
   A. Bubble gum B. Vanilla C. Smoke D. Spice

9. _____ How do additives help preserve food?
   A. Stop bacteria growth B. Improve flavor C. Change color D. They don't

10. _____ Which of the following helps bread to be fluffy?
    A. Water B. Butter C. Yeast D. Salt

Challenge questions

Choose the best answer for each question or statement (20 points each).

1. _____ Which chemical is the name for table sugar?
   A. Dextrose   B. Fructose   C. Glucose   D. Sucrose

2. _____ Which chemical gives tomatoes their red color?
   A. Carotene   B. Chlorophyll   C. Lycopene   D. Dextrose

3. _____ A calorimeter measures energy in food by _____ it.
   A. Burning   B. Eating   C. Weighing   D. Chopping

4. _____ Flavor is a combination of _____.
   A. Mass and volume B. Taste and smell C. Mass and color D. Taste and color

5. _____ Food additives are usually not used for _____.
   A. Color   B. Calories   C. Freshness   D. Flavor

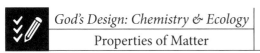

Properties of Matter

Fill in the blank with the correct term from below (3 points each).

Volume   Solid   Liquid   Melting
Mass   Evaporation   Freezing   Viscosity
Weight   Gas   Condensation   Sublimation

1. How much of a substance you have is its _____.
2. How much space something occupies is its _____.
3. How much gravity pulls on a mass is its _____.
4. The three states of matter are _____, _____ and _____.
5. When a liquid changes to a gas, it is called _____.
6. When a solid changes to a liquid, it is called _____.
7. When a liquid changes to a solid, it is called _____.
8. When a gas changes to a liquid, it is called _____.
9. When a solid changes directly to a gas, it is called _____.
10. The thickness of a liquid is called its _____.

Match the type of quantitative measurement with the proper tool (2 points each).

11. _____ Volume of a liquid   A. Meter stick
12. _____ Mass   B. Graduated cylinder
13. _____ Weight   C. Spring scale
14. _____ Temperature   D. Balance
15. _____ Volume of a cube   E. Thermometer

For each characteristic or statement, put E if it describes an element, C if it describes a compound, or M if it describes a mixture. Some statements have more than one answer (2 points each).

16. _____ Cannot be broken by ordinary chemical processes

17. _____ Contains two or more kinds of atoms

18. _____ Always has the same ratio of elements

19. _____ Iron

20. _____ Water

21. _____ Helium

22. _____ Air

23. _____ Seawater

24. _____ Only 92 of these occur in nature

25. _____ Almost all substances on earth are these

Identify each of the following changes as either a physical change (P) or a chemical change (C) (2 points each).

26. _____ Burning of a candle

27. _____ Rusting metal

28. _____ Freezing of water

29. _____ Crushing a graham cracker

30. _____ Combining oxygen and hydrogen to make water

Identify each characteristic as describing either a gas, a liquid, or a solid. Some statements have more than one answer (2 points each).

31. Molecules are far apart.

32. Has a definite shape.

33. Easily compressed.

34. Takes on the shape of its container.

35. Molecules are very close together.

Short answer (4 points each):

36. If a liquid is cooled, will it be able to dissolve more or fewer solids? _____

37. If a soft drink is very bubbly-looking, is it more likely to be warm or cold? _____

38. What similar processes are required to produce the flavors of vanilla and chocolate? _____

39. How can you tell if a solution is saturated? _____

40. How can you tell if a liquid mixture is a solution or a suspension? _____

Challenge questions

Identify each statement as origins or operational science (5 points each).

1. _____ All birds had a common reptile ancestor.
2. _____ Horses give birth to horses.

Short answer (5 points per blank):

3. List three different scales used to measure different kinds of storms. _____
   _____ _____

Mark each statement as either True or False (5 points each).

4. _____ An object is denser than another object if it has a greater volume.
5. _____ Rubbing alcohol is buoyant in water.
6. _____ Solid water is less dense than liquid water.
7. _____ Dissolving salt in water is a physical change.
8. _____ During diffusion, molecules move from an area of higher concentration to an area of lower concentration.
9. _____ A native mineral has two kinds of elements in it.
10. _____ Chromatography uses paper to separate substances in a mixture.
11. _____ Enzymes are used to harden cheese curds.
12. _____ A centrifuge uses evaporation to separate substances in a mixture.

Match each term with its definition (5 points each).

13. _____ Amount of salt in a solution.   A. Antioxidant
14. _____ Water containing calcium and magnesium. B. MSG
15. _____ Energy to raise 1g of water 1 degree C. C. Salinity
16. _____ 1000 calories (kilocalorie).   D. Hard water
17. _____ Flavor enhancer.   E. Food calorie
18. _____ Prevents reacting with oxygen.   F. Metric calorie

Ecological Quizzes and Final Exam

for Use with

Properties of Ecosystems

(*God's Design: Chemistry & Ecology*)

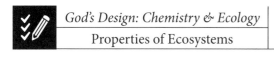

Introduction to Ecosystems

Match the term with its definition (5 points each).

1. _____ Decomposer
2. _____ Biotic
3. _____ Ecology
4. _____ Omnivore
5. _____ Abiotic
6. _____ Ecosystem/biome
7. _____ Carnivore
8. _____ Biosphere
9. _____ Flora
10. _____ Niche
11. _____ Fauna
12. _____ Habitat
13. _____ Herbivore
14. _____ Scavenger

A. The environment in which an organism lives
B. Study of the environment of living things
C. Area of the earth containing life
D. Living
E. Nonliving
F. Community of living things
G. Plants in an ecosystem
H. Animals in an ecosystem
I. Role of an organism within its environment
J. Organism that eats only plants
K. Organism that eats only animals
L. Organism that eats both plants and animals
M. Organism that eats dead plants or animals
N. Organism that breaks down dead material

Short answer (10 points each):

15. Draw a food chain with at least three levels. Label the role of each organism (producer, consumer, etc.).

16. Draw a food web with at least six organisms.

Describe each of the following relationships (5 points each).

17. Mutualism: _____

18. Parasitism: _____

Challenge questions

Short answer (20 points each):

1. Why do animals generally not migrate from one biogeographic realm to another?

2. Describe the niche of a butterfly.

3. What is carrying capacity?

4. How does the number of first order consumers in a given area compare to the number of second order consumers?

5. What law makes the oxygen, water, and nitrogen cycles necessary?

| God's Design: Chemistry & Ecology — Properties of Ecosystems | Quiz 2 | Scope: Lessons 7–11 | Total score: ____ of 100 | Name |

Grasslands & Forests

Mark each statement as either True or False (3 points each).

1. _____ A puddle of water could be considered an ecosystem.
2. _____ The amount of sunlight hitting the earth is affected by the tilt of the earth.
3. _____ Polar regions are near the equator.
4. _____ It is uncommon to have a fire in a grassland.
5. _____ Grazing animals are specially designed to eat grass.
6. _____ Burrowing animals make it harder for grass to grow.
7. _____ Trees are the dominant plants in a forest.
8. _____ Temperate forests are located near the equator.
9. _____ Arboreal animals spend most of their time in trees.
10. _____ Epiphytes are animals that live on the forest floor.
11. _____ Rainforests receive over 80 inches of rain each year.
12. _____ Pampas grass is very short.

Short answer (4 points for each blank):

13. Give three different names for grassland. _____ _____ _____

14. List the six different layers of a forest.

    A. _____   C. _____   E. _____

    B. _____   D. _____   F. _____

15. List two types of trees that you are likely to find in a deciduous forest.

    _____   _____

16. List two types of trees that you are likely to find in a coniferous forest.

    _____   _____

17. Name three common products that originally came from the tropical rainforest.

   _____   _____   _____

Challenge questions

Short answer (10 points each answer):

1. Describe how succession might take place in a forest that was destroyed by a wildfire.

   _____

   _____

2. Explain how God designed grazing animals to survive in a grassland ecosystem.

   _____

   _____

3. Explain the purpose of each of the following parts of a tree.

   a. Outer bark _____

   _____

   b. Phloem/inner bark _____

   _____

   c. Cambium _____

   _____

   d. Xylem/sapwood _____

   _____

   e. Heartwood _____

   _____

4. Place the following ecosystems in order from least amount of rainfall to greatest amount of rainfall.

   Deciduous Forest   Tropical Rainforest   Grassland   Coniferous Forest

   _____

5. Explain the importance of the tropical rainforests on the medical field.

   _____

6. Describe what you might find in an ecotone between a grassland and a forest.

   _____

   _____

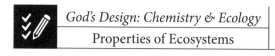

God's Design: Chemistry & Ecology — Properties of Ecosystems | Quiz 3 | Scope: Lessons 12–17 | Total score: ___ of 100 | Name

Aquatic Ecosystems

Fill in the blank with the correct term from below. Not all terms are used (4 points each).

Phytoplankton	Plankton	Inter-tidal zone	Algae bloom
Zooplankton	Atoll	Estuary	Riparian zone
Sunlit/euphotic zone	Fringing reef	Lake	Nekton
Benthos	Barrier reef	Pond	Twilight/disphotic zone
Midnight/aphotic zone	Beach	Overturn	Tributary

1. _____ are microscopic aquatic organisms that perform photosynthesis.
2. Plants and animals that live on the ocean floor are called _____.
3. A(n) _____ is a coral reef formed around a sunken volcano.
4. _____ are animals that freely move throughout the ocean.
5. Where the ocean meets the land is called a(n) _____.
6. An ecosystem where fresh water flows into the ocean is called a(n) _____.
7. Sudden rapid growth of algae is called a(n) _____.
8. Land along the banks of a river or stream is the _____.
9. A(n) _____ is a smaller stream or river that flows into a larger stream or river.
10. The _____ is the part of the shore that is covered with water at high tide and uncovered at low tide.
11. A lake that is too shallow to have an aphotic zone is referred to as a(n) _____.
12. A coral reef attached to land is a(n) _____.
13. _____ are plants and animals that move with the ocean currents.
14. The layer of water that sunlight is able to penetrate is the _____.
15. _____ is the rapid exchange of cold- and warm-water regions within a lake.

Short answer (8 points each):

16. Briefly explain why you can expect to find more varieties of plants on a rocky beach than on a sandy beach. ___

17. Why do coral grow only in relatively shallow water? _____

18. What is overturn in a lake? _____

19. Why is an estuary a very productive ecosystem? _____

20. Which organisms form the base of the food chain in the ocean? _____

Challenge questions

Mark each statement as either True or False (5 points each).

1. _____ Bioluminescent creatures produce light through a chemical reaction.
2. _____ Coral bleaching occurs when there is too much bleach in the water.
3. _____ Coral bleaching always causes the coral to die.
4. _____ Algae and coral have a symbiotic relationship.
5. _____ A dune system is an example of ecological succession.
6. _____ A maritime forest usually has large trees.
7. _____ The grass in a dune system helps to stabilize the dunes.
8. _____ Dune grass must be tolerant to salt and wind.
9. _____ Land can only be part of a single watershed.
10. _____ The Mississippi River Basin is the largest watershed in the United States.
11. _____ Water from the Mississippi River mixes quickly with the Gulf of Mexico.
12. _____ The Great Lakes provide water and work for over 35 million people.
13. _____ Invasive species are not a real threat to animals in the Great Lakes.
14. _____ The Great Lakes can generate their own weather systems.
15. _____ The Mississippi River has the largest volume of any river.
16. _____ The Nile River is one of the longest rivers in the world.
17. _____ River ecosystems vary as the speed of the river changes.
18. _____ The Volga River is an important ecosystem in Russia.
19. _____ The Amazon River has the largest watershed in the world.
20. _____ Bioluminescent creatures live primarily in the sunlit zone.

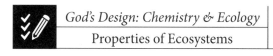

Extreme Ecosystems

1. Place the animals below in the ecosystem(s) you are likely to find them **(40 points)**.

   Arctic fox   Scorpion   Snake   Toad
   Bat   Ground squirrel Mountain lion   Camel
   Reindeer   Lizard   Big horned sheep   Crayfish
   Moose   Canada goose   Cricket

Tundra	Desert	Oasis	Mountain	Cave

Choose the best answer for each statement or question (4 points each).

2. _____ What is the layer of permanently frozen ground in the tundra called?
   A. Swamp   B. Permafrost   C. Chill layer   D. Glacier

3. _____ Which do not help plants survive in the tundra?
   A. Large leaves   B. Short stems   C. Fuzzy leaves   D. Close groupings

4. _____ On average, how much moisture does the tundra receive each year?
   A. 0–10 inches   B. 10–20 inches   C. 20–30 inches   D. 30–60 inches

5. _____ On average, how much moisture does a desert receive each year?
   A. 0–10 inches   B. 10–20 inches   C. 20–30 inches   D. 30–60 inches

6. _____ Which of the following helps animals survive in the desert?
   A. Long hair   B. Daytime activity C. Short lifecycle   D. Large body

7. _____ Which are you not likely to find in an oasis?
   A. Water   B. Palm trees   C. Grass   D. Sharks

8. _____ You would expect the temperature in an oasis to be _____ than in the desert.
   A. Hotter   B. Cooler   C. Dryer   D. The same as

9. _____ Which ecosystem would not likely be found on a mountain?
   A. Tundra   B. Deciduous forest C. Grassland   D. Desert

10. _____ What is the point above which no trees will grow?
   A. Timberline   B. Rockline   C. Grassline   D. Continental Divide

11. _____ Which is likely to increase as you go up a mountain?
   A. Oxygen   B. Temperature   C. Amount of snow   D. Gravity

12. _____ Which animal are you likely to find only in Australian chaparral?
   A. Koala   B. Woodrat   C. Badger   D. Squirrel

13. _____ Which condition does not contribute to fire in the chaparral?
   A. Heat   B. Dryness   C. Rain   D. Lightning

14. _____ Which animal is most important to cave ecosystems?
   A. Scorpions   B. Shrimp   C. Fish   D. Bats

15. _____ Which of the following does not describe an animal that visits or lives in a cave?
   A. Troglobites   B. Troglophiles   C. Troglodytes   D. Trogloxenes

16. _____ Which sense is least useful inside a cave?
   A. Hearing   B. Sight   C. Smell   D. Touch

Challenge questions

Short answer (5 points per blank):

1. List three ways that polar bears are designed to live in the tundra. _____

2. The largest hot desert in the world is the _____

3. Where is this desert located? _____

4. How has this desert changed since the time of the Genesis Flood?

5. List three major products that come from deserts. _____

6. What is the tallest mountain in the Himalayas? _____

7. Name three animals found only in the Himalayas. _____

8. List three possible fire cues for seed germination. _____

9. What method do insect-eating bats use to find their food? _____

10. List three different kinds of food eaten by different kinds of bats. _____

| *God's Design: Chemistry & Ecology* **Properties of Ecosystems** | Quiz 5 | Scope: Lessons 24–27 | Total score: ____ of 100 | Name |

Animal Behaviors

Mark each statement as either True or False (4 points each).

1. _____ Hibernation is a seasonal behavior for animals.
2. _____ An animal's heartbeat is higher during hibernation.
3. _____ Estivation occurs during the winter.
4. _____ Butterflies often migrate hundreds of miles.
5. _____ Migrating birds often fly in a V formation.
6. _____ Trickery is most animals' first defense.
7. _____ Most animals are defenseless against their enemies.
8. _____ Camouflage is a good animal defense.
9. _____ Prairie dogs can alert their colony to possible dangers.
10. _____ Adaptation can be a physical characteristic or a behavior.
11. _____ Natural selection does not really occur.
12. _____ Animals can adapt because their DNA allows for great variety.
13. _____ Man is needed to maintain a balance in most ecosystems.
14. _____ When the predator population increases, the prey population decreases.
15. _____ God designed territoriality as a way to control populations of animals.

Short answer (8 points each):

16. Describe how territoriality helps control animal populations. _____

17. Explain why bears do not truly hibernate. _____

18. Describe one type of animal defense. _____

19. Explain how adaptations are a result of creation and not evolution. _____

20. What is the most likely trigger for seasonal behaviors? _____

_____

Challenge questions

Match the term with its definition (10 points each).

1. _____ Plant defense against animals.   A. Darwin's finches

2. _____ Plant defense against weather.   B. Thorns

3. _____ Species developing from a common ancestor.   C. Natural selection

4. _____ Birds commonly used to support evolution.   D. Bugling

5. _____ Ability of a species to survive better than others.   E. DDT

6. _____ Organisms with human-modified DNA.   F. Dormancy

7. _____ Chemicals used to kill unwanted animals.   G. GMO

8. _____ Chemical used to control malaria.   H. Pesticides

9. _____ Display done to attract a mate.   I. Adaptive radiation

10. _____ Noise made by male elk to attract a mate.   J. Animal courtship

| God's Design: Chemistry & Ecology — Properties of Ecosystems | Quiz 6 | Scope: Lessons 28–33 | Total score: ____ of 100 | Name |

Ecology & Conservation

Identify each of the following as either natural (N), man-made (M), both (B), or unknown (U) in origin (4 points each).

1. _____ Cooler weather due to ash in the atmosphere from a volcanic eruption
2. _____ Smoke in the air from a fire started by a lightning strike
3. _____ Acid rain
4. _____ Species extinction
5. _____ Habitat reduction
6. _____ Invasive species introduction
7. _____ Captive breeding
8. _____ Air pollution
9. _____ Water pollution
10. _____ Greenhouse effect
11. _____ Global warming
12. _____ Buffering capacity
13. _____ Deforestation
14. _____ Plastic recycling
15. _____ Overhunting

Fill in the blank with the correct term (2 points each blank).

16. The 3 R's of conservation are _____, _____, and _____.

17. Ranch animals must share _____, _____, and _____ with native animals.

18. Species that are no longer alive are said to be _____.

19. The zebra mussel is considered a(n) _____ species.

20. The most costly captive breeding program was to save the _____.

21. Three main areas of pollution include _____, _____, and _____.

22. _____ is caused by sulfur dioxide and nitrogen oxides in the air.

23. Materials that naturally decompose are said to be _____.

24. An area's ability to neutralize acid rain is called its _____.

25. One thing I can do to help the environment is _____.

26. List four ways that people impact the environment. _____, _____, _____, _____

Challenge questions

Short answer (10 points each blank):

1. Briefly describe a biblical view of ecology.
   _____

2. List two government organizations that are committed to protecting the environment.
   _____   _____

3. Briefly explain how ozone depletion can occur.
   _____

4. List two alternative sources of energy that could replace fossil fuels.
   _____   _____

5. Why are fossil fuels considered non-renewable resources?
   _____
   _____

6. Explain two possible problems with plastic recycling.
   _____
   _____

7. Describe your plan for how you will be a good steward of God's environment.
   _____
   _____

Properties of Ecosystems

Match the term with its definition (2 points each).

1. _____ Flow of energy from one organism to another.   A. Habitat
2. _____ Area of the earth containing life.   B. Biosphere
3. _____ Organisms that produce food.   C. Food chain
4. _____ Non-living.   D. Producers
5. _____ The environment in which an organism lives.   E. Abiotic
6. _____ Ability of the soil to neutralize acid.   F. Steppe
7. _____ Top layer of a forest.   G. Buffering capacity
8. _____ Living life primarily in trees.   H. Emergent layer
9. _____ Grassland of Europe and Asia.   I. Arboreal
10. _____ Microscopic aquatic animals.   J. Zooplankton
11. _____ Where fresh water flows into the ocean.   K. Cave
12. _____ Land along the banks of a river or stream.   L. Estivate
13. _____ Coral reef formed around a sunken volcano.   M. Riparian zone
14. _____ Deep sleep during the summer.   N. Atoll
15. _____ Cavern in a mountain or underground.   O. Estuary

Short answer:

16. Draw a picture of a food chain with at least three links **(3 points)**.

17. Draw a picture of a food web with at least six different and interconnected organisms **(3 points)**.

18. Draw a picture of the water cycle **(4 points)**.

19. Put the organisms below in the ecosystem in which they are most likely to be found **(15 points)**.

Plants		Animals		
Plankton	Heather	Zebra	Shark	Millipede
Lemon tree	Algae	Gazelle	Coyote	Bat
Sage	Pineapple	Polar bear	Monkey	Tree frog
Cocoa tree	Ephemerals	Caribou	Ptarmigan	Toucan
Grass		Jellyfish	Octopus	Crab
		Capybara	Blind fish	Bison

Tundra	Grassland	Rainforest	Ocean	Cave

20. List at least three characteristics of each of the following ecosystems **(15 points)**.

Coral reef	Estuary	Deciduous forest	Desert	Mountain

Mark each statement as either True or False (2 points each).

21. _____ Many different animals exhibit seasonal behaviors.

22. _____ Bears do not truly hibernate since they can wake up in the winter.

23. _____ Monarch butterflies complete their migration in one generation.

24. _____ Female birds are the ones that usually sing and defend their territory.

25. _____ All animals were originally designed to eat plants.

26. _____ Trickery is the first instinct most animals have for defending themselves.

27. _____ Animals can use claws and teeth for defense.

28. _____ Territoriality is the primary way that populations are controlled today.

29. _____ Territoriality is the original way God designed population control.

30. _____ Natural selection does not really occur.

31. _____ Pollution can sometimes have natural causes.

32. _____ Recycling is one way to help reduce man's impact on nature.

33. _____ We should all panic about global warming.

34. _____ Man is the only reason species become extinct.

35. _____ People have made great progress in reducing acid rain.

Challenge questions

Fill in the blank with the correct term (5 points each).

1. The maximum population an area can support is called its _____.
2. A(n) _____ is a transitional area between two ecosystems.
3. _____ plants are the first plants to move into an area after a natural disaster.
4. The final, stable ecosystem of a succession is called the _____ ecosystem.
5. _____ is a relationship between two organisms in which both are benefited.
6. The land drained by a particular body of water is called a(n) _____.
7. _____ is the use of sound waves by bats to detect objects.
8. Behavior performed to attract a mate is called _____.
9. _____ refers to several species that develop from a common ancestor.
10. Ozone protects the earth from _____.
11. _____ are the molecules that are believed to cause ozone destruction.
12. Plastics are made from molecules called _____.
13. Resources that cannot be replaced are _____.

Short answer:

14. Draw a population pyramid for an aquatic ecosystem **(6 points)**.

15. Describe a possible succession for an area of pine forest after a forest fire **(6 points)**.

   _____
   _____

16. Draw a cross section of a tree trunk. Label all parts **(6 points)**.

17. List at least three commercial products from the desert **(6 points)**. _____

   _____ _____

18. Explain how some plants can germinate only when there is a fire **(6 points)**.

   _____
   _____

19. Describe one animal courtship ritual you found interesting **(5 points)**.

   _____
   _____

Atomic & Molecular Quizzes and Final Exam
for Use with
Properties of Atoms & Molecules
(*God's Design: Chemistry & Ecology*)

Atoms & Molecules

Use the periodic table of the elements from your book to answer the following questions (10 points each blank).

Label the parts of a helium atom.

A. _____

B. _____

C. _____

B and C together form the _____.

Match the term with its definition (6 points each).

1. _____ Anything that has mass and takes up space.
2. _____ A positively charged particle in an atom.
3. _____ A negatively charged particle in an atom.
4. _____ A neutral particle in the nucleus.
5. _____ Mass of a proton or neutron.
6. _____ Compact center of the atom.
7. _____ Two atoms of the same element connected together.
8. _____ Part of matter that cannot be broken down chemically.
9. _____ Number of protons an element has.
10. _____ Two or more atoms chemically bonded.

A. Atomic mass unit
B. Neutron
C. Molecule
D. Matter
E. Proton
F. Atom
G. Diatomic molecule
H. Atomic number
I. Electron
J. Nucleus

Challenge questions

Short answer:

1. Refer to the periodic table of the elements to complete the following chart **(4 points each box)**.

Element	Atomic #	Atomic mass	# of protons	# of electrons	Most common # of neutrons
Carbon					
Aluminum					
Tungsten					

2. Based on the electron configurations for the following elements, which would not be likely to bond with any other elements? **(10 points)** _____

    Potassium (K)　　　　Nitrogen (N)　　　　Argon (Ar)

3. What is an isotope? **(10 points)** _____
4. What is a valence electron? **(10 points)** _____
5. What is the electron configuration for silicon (Si)? **(10 points)** _____

| God's Design: Chemistry & Ecology Properties of Atoms & Molecules | Quiz 2 | Scope: Lessons 5–10 | Total score: ____of 100 | Name |

Elements

Short answer (5 points each):

1. What do elements in a column of the periodic table have in common?
   _____

2. What do elements in a row of the periodic table have in common?
   _____

3. Which column of elements is most stable? _____

4. Elements in which column are most likely to react with elements in column VIA? _____

5. Elements in which column are most likely to react with elements in column VIIA? _____

Write metal, metalloid, or nonmetal to match the type of element to its characteristics (5 points each).

6. _____ Silvery luster

7. _____ Ductile

8. _____ Conducts electricity

9. _____ Does not conduct electricity

10. _____ Solid at room temperature

11. _____ Not shiny

12. _____ Somewhat malleable

13. _____ Most often a gas

14. _____ Semiconductor

15. _____ Malleable

Mark each statement as either True or False (5 points each).

16. _____ Hydrogen is very reactive.

17. _____ Oxygen is lighter than hydrogen.

18. _____ Hydrogen is the most common element on earth.

19. _____ All elements are recycled; they are not destroyed.

20. _____ Carbon forms organic compounds.

Challenge questions

Match the term with its definition (10 points each).

1. _____ Column of the periodic table.
2. _____ Row of the periodic table.
3. _____ Metals in column 1.
4. _____ Metals in column 2.
5. _____ Man-made elements.
6. _____ Ball-shaped carbon molecule.
7. _____ Thread-like cylinders of carbon atoms.
8. _____ Technology that combines hydrogen and oxygen to produce electricity.
9. _____ Atoms of the same element that link together in different ways.
10. _____ Elements that do not occur naturally.

A. Alkali metals
B. Family
C. Hydrogen fuel cell
D. Alkali-earth metals
E. Synthetic elements
F. Period
G. Buckyball
H. Nanotubes
I. Synthetic elements
J. Allotropes

Bonding

For each characteristic below, write I if it describes an ionic substance, C for a covalent substance, and M for a metallic substance. Some characteristics have more than one answer (5 points each).

1. _____ Formed by elements with very different levels of electronegativity
2. _____ High melting point
3. _____ Electrons are shared between two or three atoms
4. _____ Insoluble in water
5. _____ Forms ions
6. _____ Electrons are given up or pulled away
7. _____ Does not conduct electricity
8. _____ Sharing of electrons on a large scale
9. _____ Conducts electricity
10. _____ Flexible

Short answer (10 points each):

11. How are crystals formed? _____
12. What is the smooth side of a crystal called? _____
13. What process is necessary for ceramics to become strong? _____
14. What is the common ingredient in all natural ceramics? _____
15. Name three traditional ceramics. _____

Challenge questions

Mark each statement as either True or False (6 points each).

1. _____ Ionic bonding occurs between elements with very different electronegativities.
2. _____ Ions are electrically neutral.
3. _____ Ionic bonds occur between nonmetals.
4. _____ Sodium fluoride is an ionic compound.
5. _____ Covalent compounds easily conduct electricity.
6. _____ Covalent bonds occur between nonmetals.
7. _____ Metallic materials easily dissolve in water.
8. _____ Metallic bonds have free electrons.
9. _____ Metallic bonds form between elements with similar low electronegativities.
10. _____ Brass is an alloy of copper and zinc.
11. _____ Steel is an alloy of copper and tin.
12. _____ Hydrates usually feel wet.
13. _____ Hydrates can help prevent the spread of fire.
14. _____ Resorbable ceramics are absorbed into the body.
15. _____ Inert ceramics react with the body.

Short answer (5 points each):

16. What is the difference between a cation and an anion? _____

_____

17. Heating will usually remove water from hydrates. What is this process called? _____

_____

Chemical Reactions

Mark each statement as either True or False (6 points each).

1. _____ All chemical reactions are fast.
2. _____ A catalyst speeds up a chemical reaction.
3. _____ Endothermic reactions use up heat.
4. _____ A fireworks explosion is an endothermic reaction.
5. _____ The same number of atoms must appear on both sides of a chemical equation.
6. _____ Chemical equations demonstrate the first law of thermodynamics.
7. _____ Reactants are on the left side of a chemical equation.
8. _____ Catalysts are used up in a chemical reaction.
9. _____ Inhibitors speed up a chemical reaction.
10. _____ Sometimes inhibitors are helpful.
11. _____ Catalysts lower the energy required for a chemical reaction to occur.
12. _____ Exothermic reactions release energy.
13. _____ The product of an exothermic reaction is cooler than the reactants.
14. _____ Chemical reactions are rare.
15. _____ Heat can increase the reaction rate.

Short answer (5 points each):

16. What happens in a composition reaction? _____

17. What happens in a decomposition reaction? _____

Challenge questions

Identify each of the following reactions as composition (c), decomposition (d), single displacement (sd), or double displacement (dd) (6 points each).

1. _____ $H_2SO_4 + 2\ LiOH \longrightarrow Li_2SO_4 + 2\ H_2O$

2. _____ $P_4 + 10\ Cl_2 \longrightarrow 4\ PCl_5$

3. _____ $CO_2 \longrightarrow C + O_2$

4. _____ $2\ Na + 2\ H_2O \longrightarrow H_2 + 2\ NaOH$

5. _____ $AgNO_3 + HCl \longrightarrow AgCl + HNO_3$

Short answer (7 points each blank):

6. List three ways to increase the reaction rate of a chemical reaction.

   _____

   _____

   _____

7. List two groups of catalysts. _____ _____

8. Which type of catalyst is found in a catalytic converter? _____

9. Which type of catalyst will bond with a reactant? _____

10. What is the name for the energy stored in chemical bonds? _____

11. If the enthalpy of the products of a reaction is higher than the enthalpy of the reactants, is the reaction endothermic or exothermic?

    _____

    _____

12. What is the first law of thermodynamics?

    _____

    _____

God's Design: Chemistry & Ecology	Quiz 5	Scope: Lessons 21–24	Total score: ____ of 100	Name
Properties of Atoms & Molecules				

Acids & Bases

Choose the best answer for each question (10 points each).

1. _____ Which is not a type of chemical analysis?
   A. Flame test   B. Spectrometer   C. Indicator   D. Temperature

2. _____ pH indicators can tell the strength of which type of compound?
   A. Acid   B. Salt   C. Water   D. Tea

3. _____ What flower can indicate the pH of the soil by the color of its flowers?
   A. Rose   B. Hydrangea   C. Tulip   D. Daisy

4. _____ Which of the following is not an acid?
   A. Orange juice   B. Vinegar   C. Ammonia   D. Stomach fluid

5. _____ Which of the following is not a base?
   A. Saliva   B. Milk of magnesia   C. Soap   D. Lye

6. _____ What is formed when an acid combines with a base?
   A. Salt   B. Hydroxide   C. Hydronium   D. Balloons

7. _____ Which is not a characteristic of acids?
   A. Sour taste   B. Slippery   C. Corrosive   D. Conducts electricity

8. _____ Which is not a characteristic of bases?
   A. Sour taste   B. Slippery   C. Corrosive   D. Conducts electricity

9. _____ Which acid is the most produced chemical in the United States?
   A. Sulfuric   B. Hydrochloric   C. Formic   D. Ascetic

10. _____ What common product is made primarily from salts?
    A. Cake   B. Taffy   C. Fertilizer   D. Batteries

Challenge questions

Choose the best answer for each statement (20 points each).

1. _____ Electroplating is depositing a thin layer of metal on a(n) _____.
   A. Battery B. Base C. Conductor D. Plate

2. _____ Titration allows you to calculate how many _____ of acid or base are in an unknown sample.
   A. Molecules B. Electrons C. Protons D. Neutrons

3. _____ A proton donor is another name for a(n) _____.
   A. Electron B. Acid C. Base D. Salt

4. _____ A proton acceptor is another name for a(n) _____.
   A. Electron B. Acid C. Base D. Salt

5. _____ You can identify the acid in a chemical equation because it loses a(n) _____ atom.
   A. Oxygen B. Nitrogen C. Helium D. Hydrogen

God's Design: Chemistry & Ecology
Properties of Atoms & Molecules

Quiz 6 | Scope: Lessons 25–28 | Total score: ____ of 100 | Name _____

Biochemistry

Short answer (4 points each blank):

1. Identify two chemical reactions that sustain life. _____, _____
2. Name three main chemical compounds in food. _____
   _____, _____
3. What type of catalyst increases the rate of digestion processes? _____
4. What is an animal called that eats dead animals? _____
5. Name two types of decomposers. _____, _____
6. Name an element that is recycled by decomposers. _____
7. Name three ways to keep farmland productive. _____
   _____, _____
8. Who was the discoverer of penicillin? _____
9. What is the most common solvent in the human body? _____

Match the term with its definition (5 points each).

10. _____ Kills unwanted insects.
11. _____ Kills unwanted plants.
12. _____ Kills unwanted fungus.
13. _____ Farming without man-made chemicals.
14. _____ Growing plants without soil.
15. _____ Medicine to kill bacteria.
16. _____ Medicine to encourage natural defenses.
17. _____ Some of these plants have natural medicinal value.

A. Hydroponics
B. Insecticide
C. Organic
D. Herbicide
E. Antibiotics
F. Vaccine
G. Herbs
H. Fungicide

Challenge questions

Short answer (25 points each):

1. What are two conditions that inhibit enzyme reaction rate?

   _____

   _____

2. What condition most promotes decomposition?

   _____

3. What are two controversies surrounding organic farming?

   _____

   _____

4. Briefly explain how chemotherapy works to treat cancer.

   _____

   _____

| God's Design: Chemistry & Ecology | Quiz 7 | Scope: Lessons 29–33 | Total score: ____ of 100 | Name |

Applications of Chemistry

Briefly explain how chemistry is used in the making of each of the following items (8 points each).

1. Perfume _____
2. Rubber _____
3. Plastic _____
4. Fireworks _____
5. Rocket fuel _____

Mark each statement as either True or False (4 points each).

6. _____ Vulcanization makes rubber useful in most temperatures.
7. _____ Rubber is made from cellulose.
8. _____ A polymer is a very short molecule.
9. _____ Today, synthetic rubber is more widely used than natural rubber.
10. _____ Perfume smells the same in the bottle as on your skin.
11. _____ Latex is a natural polymer.
12. _____ Bakelite was the first useful plastic.
13. _____ Plastic is an important product in American life.
14. _____ Fireworks are different colors because of different chemical compounds used.
15. _____ Recipes for fireworks are freely shared.
16. _____ Kerosene and carbon dioxide are common rocket fuels today.
17. _____ Newton's third law of motion is important in rocket design.
18. _____ Combustion is a chemical reaction that produces large amounts of heat.
19. _____ Flower-scented perfume is made from the oil in flower petals.
20. _____ Plastic is made from latex.

Challenge questions

Mark each statement as either True or False (10 points each).

1. _____ Scents smell the same on every person.
2. _____ Silk is a natural polymer.
3. _____ A milk protein can be used as a glue.
4. _____ Creating polymers is very difficult.
5. _____ Flames are always the same color.
6. _____ Sodium chloride burns with a yellow flame.
7. _____ Hypergolic rocket fuel is not very common.
8. _____ Solid rocket engines must use up all of their fuel once they are ignited.
9. _____ Liquid rocket fuel is used in most space rockets.
10. _____ It is harder to control the rate at which cryogenic fuel burns than the rate at which solid rocket fuel burns.

Atoms & Molecules

For each pair of elements, write I if they are most likely to form an ionic bond, C for covalent bond, or M for metallic bond (4 points each).

1. _____ Na + Cl
2. _____ H$_2$ + O
3. _____ O + O
4. _____ K + Br
5. _____ Al + Al
6. _____ Mg + O
7. _____ C + O$_2$
8. _____ Ag + Ag
9. _____ Cu + Cu

Fill in the blanks with the terms from below (3 points each).

Catalyst	Fat	Carbohydrate	Exothermic
Enzyme	Endothermic	Protein	
Base	Salt	Acid	

10. A(n) _____ can be used to speed up a chemical reaction.

11. The products of a(n) _____ reaction have a higher temperature than the reactants.

12. The products of a(n) _____ reaction have a lower temperature than the reactants.

13. A(n) _____ is a catalyst that increases the rate of digestion.

14. An acid and a base combine to form a(n) _____.

15. A substance is a(n) _____ if it releases H$^+$ ions when dissolved in water.

16. A substance is a(n) _____ if it releases OH$^-$ ions when dissolved in water.

Draw and label a model of a helium atom, which has an atomic number of 2 and an atomic mass of 4 (6 points).

Choose one of the following topics and briefly explain how chemistry plays a role in it (7 points).

Farming   Medicine   The nitrogen cycle

_____

_____

Match the term with its definition (3 points each).

17. _____ What natural rubber is made from.   A. Vulcanization

18. _____ What synthetic rubber is made from.   B. Latex

19. _____ A long flexible chain of molecules.   C. Combustion

20. _____ Process that makes rubber strong and flexible.   D. Petroleum

21. _____ A natural polymer found in plants.   E. Polymer

22. _____ Process of burning that releases large amounts of heat.   F. Cellulose

Short answer (4 points each):

23. List three characteristics of a metal.

_____

_____

24. List three characteristics of a nonmetal.

_____

_____

25. Explain the chemical reaction involved in your favorite experiment from this book.

_____

_____

_____

Challenge questions

1. Use the periodic table of the elements to complete the following chart **(28 points – 1 for each box)**.

Element	Symbol	Atomic #	Atomic mass	# electrons	# protons	Most likely # neutrons
	Fe					
Potassium						
			80			
					36	

Fill in the blanks with the terms from below (6 points each).

Alkali metals	Transition metals	Inert ceramic	Proton donor
Alkali-earth metals	Hydrates	Heterogeneous catalyst	Proton acceptor
Noble gases	Resorbable ceramic	Homogeneous catalyst	Reaction rate

2. Temperature can increase the _____ of a chemical reaction.
3. The elements in the first column of the periodic table are _____.
4. The elements in the last column of the periodic table are _____.
5. The elements in the center of the periodic table are _____.
6. The elements in the second column of the periodic table are _____.
7. A bioceramic that does not react with the body is a(n) _____.
8. A bioceramic that dissolves in the body is a(n) _____.
9. An acid is a(n) _____.
10. A base is a(n) _____.
11. Molecules that have water bonded to them are _____.
12. A(n) _____ is in the same phase as the reactants.
13. A(n) _____ is in a different phase from the reactants.

Worksheet Answer Keys

for Use with

God's Design: Chemistry & Ecology

Properties of Matter — Worksheet Answer Keys

1. Introduction to Experimental Science

What did we learn?

1. What is matter? **Anything that has mass and takes up space.**
2. What do chemists study? **They study the way matter reacts with other matter and the environment.**
3. What is an experiment? **A controlled test.**

Taking it further

1. Why is it important to study chemistry? **Chemistry is important to every other area of science.**
2. What are two things you need to know before conducting an experiment? **The purpose and what you expect to happen.**

2. The Scientific Method

What did we learn?

1. What is the overall job of a scientist? **To systematically study the physical world.**
2. What are some areas that cannot be studied by science? **Morality, religion, philosophy, history.**
3. What are the five steps of the scientific method? **Learn or observe, ask a question, make a hypothesis, design and perform a test, check the results, and draw conclusions.**

Taking it further

1. Why was it necessary to have bottle number 1 in the experiment? **Bottle 1 had only water and yeast. This is called a control. It shows how much gas was produced without a sweetener, so you can tell exactly how much gas was caused by adding the sugar and molasses in the other bottles.**
2. What other sweeteners could you try in your experiment? **Honey, corn syrup, fruit juice.**
3. What sweeteners were used in the bread at your house? **Look at the ingredients list on the package if you do not bake your own bread. Possible answers are sugar, corn syrup, and honey.**
4. Why do you think that sweetener was used? **Reasons vary, but amount of gas produced, cost, color, and taste are all important factors in why companies use the ingredients they do.**

Scientific Method Worksheet

Taste, color, and texture are all affected by the sweetener used, so even if molasses produces the most gas, you may not like the way it makes your bread taste or look.

3. Tools of Science

What did we learn?

1. What is the main thing a scientist does as he/she studies the physical world? **Makes observations.**
2. What are the two types of observations that a scientist can make? **Qualitative observations are ones made by the 5 senses without numerical data. Quantitative measurements or observations are made using instruments that generate numerical or other objective data.**
3. What is the main problem with qualitative observations? **The observations may vary from person to person because we each perceive things differently.**
4. What are some scientific tools used for quantitative observations? **Balance, graduated cylinder, thermometer, meter stick, spectrometer, etc.**

Taking it further

1. What qualitative observations might you make when observing the experiment in lesson 1? **You might observe that the metal spoon is hotter than a wooden spoon or that butter begins to melt faster or slower on certain items.**
2. What quantitative observations might you make when observing the experiment in lesson 1? **You might measure the temperature of the water and the temperature of each item. You did measure the length of time it took for the butter to begin to melt on each item. You could also measure the length of time it takes for the butter to completely melt on each item.**

Scientific Tools Worksheet

Answers to "Conclusion" questions: Quantitative measurements are more accurate. In general, quantitative measurements are more useful; however, this depends on what you are trying to accomplish. It is not always necessary to make quantitative measurements. You may only need to know if something is warm or melted without having to measure its temperature, for example.

4. The Metric System

What did we learn?

1. What are some units used to measure length in the Old English/American measuring system? **Inch, foot, yard, mile, rod, hand, span.**

2. What is the unit used to measure length in the metric system? **Meter.**

3. What metric unit is used for measuring mass? **Gram.**

4. What metric unit is used for measuring liquid volume? **Liter.**

5. Why do scientists use the metric system instead of another measuring system? **It is easy to convert from one unit to another, and it is based on only a few basic units. In fact, liters and grams are actually based on the meter. For example, the liter is actually the volume of a cube that is .1 x .1 x .1 meters, and a gram is the mass of 1/1000 of a liter, or one cubic centimeter, of water.**

Taking it further

1. What metric unit would be best to use to measure the distance across a room? **Meters would be the best unit.**

2. What metric unit would you use to measure the distance from one town to another? **The distance would be a very large number if you used meters, so kilometers would be a better choice.**

3. What metric unit would you use to measure the width of a hair? **This is much smaller than a meter, so a millimeter or micrometer would be a better choice.**

Using Metric Units

1. 1 liter of water
2. 20,000 grams
3. 4,000 meters
4. 6 centimeters
5. 500 decigrams

5. Mass vs. Weight

What did we learn?

1. What is the difference between mass and weight? **Mass is the amount of material there is in an object, and weight is how much gravity pulls down on an object.**

2. How do you measure mass? **By using a balance to compare an object to a known mass.**

3. How do you measure weight? **By using a spring scale that is marked for known weights.**

Taking it further

1. What would your weight be in outer space? **Nearly zero because there is very little gravity in space.**

2. What would your mass be in outer space? **The same as it is on earth.**

3. Name a place in the universe where you might go to increase your weight without changing your mass. **Any of the larger planets such as Jupiter or Saturn. Of course, you cannot really go there and you could not survive there if you could, but the gravity is much higher there than on earth, so you would weigh much more there.**

Challenge: Mass & Weight Units Worksheet

1. _Weight_
2. _Mass_
3. _Mass_
4. _Weight_
5. _Weight_
6. _Mass_
7. _Weight_
8. _Mass_
9. _Mass_
10. _Mass_
11. _Mass_
12. _Weight_

6. Conservation of Mass

What did we learn?

1. What is the law of conservation of mass? **Matter cannot be created nor destroyed. It can change form, but it does not go away.**

2. How is the mass of water changed when it turns to ice? **It does not change.**

Taking it further

1. If you start with 10 grams of water and you boil it until there is no water left in the pan, what happened to the water? **The 10 grams of water turned into 10 grams of steam and entered the air, but it did not disappear or go away.**

2. Why is the law of conservation of mass important to understanding the beginning of the world? **It shows that matter cannot create itself or be created by anything in nature. Therefore, it had to be created by something outside of nature. We know from the Bible that all matter was created by God.**

Challenge: Conservation of Mass Worksheet

The mass of the bottle, liquid, and paper is less after the reaction because some of the matter turned into gas and escaped from the bottle. The missing mass is in the CO_2 molecules. Mass should not change when using the balloon, but if it does, it is likely that some gas escaped around the edge of the balloon.

7. Volume

What did we learn?

1. What is volume? **The amount of room or space something occupies.**

2. Does air have volume? **Yes, even though you can't see it, it still takes up space. It expands to fill up the available space. Think about a balloon. The air forces the balloon to expand, visibly showing how much room the air is taking up.**

Taking it further

1. If you have a cube that is 10 centimeters on each side, what would its volume be? **10 cm x 10 cm x 10 cm = 1000 cubic centimeters.**

2. Why is volume important to a scientist? **The volume of matter can be related to many things that scientists are interested in. For example, the volume that a certain amount of fuel occupies determines how a vehicle will be designed.**

Challenge: Calculating Volume Worksheet

Answers will vary. Be sure student correctly used each formula.

8. Density

What did we learn?

1. What is the definition of density? **The mass of an object divided by its volume.**

2. If two substances with the same volume have different densities, how can you tell which one is the densest? **If they have the same volume, the one that is heavier will have the higher density.**

Taking it further

1. If you have two unknown substances that both appear to be silvery colored, how can you tell if they are the same material? **Measure their densities. Platinum has a density of 21.45 g/cc, lead is 11.3 g/cc, and aluminum is 2.7 g/cc. This may give you a clue to the material's identity.**

2. If two objects have the same density and the same size, what will be true about their masses? **They will have the same mass.**

3. If you suspect that someone is trying to pass off a gold-plated bar of lead as a solid gold bar, how can you test your theory? **Measure the density of the bar. Gold has a density of 19.3 g/cc while lead has a density of 11.3 g/cc. Even though lead may seem heavy, it is not as dense as gold.**

4. Why does the ping-pong ball have a lower density than the golf ball? **It is filled with air. Air is very light compared to most substances. The golf ball is filled with plastic, rubber, or other solid materials.**

Challenge: Density Experiment Worksheet

Answers will vary. Be sure student used the correct procedure to calculate density.

9. Buoyancy

What did we learn?

1. What is buoyancy? **The ability to float.**

2. If something is buoyant, what does that tell you about its density compared to that of the substance in which it floats? **It means that the object's density is less than the density of the substance that it is floating in.**

3. Are you buoyant in water? **Probably, especially if you are holding your breath.**

Taking it further

1. What are some substances that are buoyant in water besides you? **Ivory soap, a leaf, paper, oil, etc.**

2. Based on what you observed, which is denser, water or alcohol? **Water is denser. Oil will float on the water but sinks in the alcohol.**

3. Why is a foam swimming tube or a foam life ring able to keep a person afloat in the water? **Foam is a material that has air trapped in it, so it is not very dense. Even with the person's weight/mass added to it, the foam object's density remains lower than the density of the water.**

4. Why is it important to life that ice is less dense than water? **Otherwise, rivers and lakes would freeze from the bottom up, and no life could survive in them.**

10. Physical & Chemical Properties

What did we learn?

1. What are some physical properties of matter? **These could include color, texture, temperature, density, mass, state, etc.**

2. What is a chemical change? **When two or more substances combine to form a different substance.**

3. Give an example of a chemical change. **There are innumerable examples. Some common chemical changes that might be mentioned include photosynthesis, hydrogen and oxygen combining to form water, vinegar and baking soda combining to form carbon dioxide, yeast turning sugar into carbon dioxide, rust, digesting food, etc.**

Taking it further

1. How can you determine if a change in matter is a physical change or a chemical change? **Find out if the ending matter is the same type of matter as what you started with. Chemical changes often involve release of energy such as heat, light, or sound.**

2. Diamond and quartz appear to have very similar physical properties. They are both clear crystalline substances. However, diamond is much harder than quartz. How would this affect each one's effectiveness as tips for drill bits? **The quartz-tipped drill would quickly wear down and be ineffective. Diamond-tipped drills are very hard and very effective at drilling through nearly any other substance. It is important to understand physical properties of matter as well as chemical properties.**

Challenge: Physical or Chemical Properties Worksheet

1. _P_ Liquid water becoming steam (**Steam is still water.**)
2. _C_ Flavor/taste (**Texture contributes to taste, but most of the flavor is detected through chemical reactions in the taste buds.**)
3. _C_ Burning of wood/fire (**The wood and oxygen combine to form ash, water, and carbon dioxide. Heat and light are released during this chemical reaction.**)
4. _P_ Filling a balloon with air (**The balloon changes shape but is still a balloon, and the air is still air.**)
5. _P_ Softness (**Softness is a physical characteristic of the object.**)
6. _P_ Making ice cream (**The sugar & milk mixture is frozen but does not become something else.**)
7. _C_ Digesting food (**Some physical aspects but is primarily chemical.**)
8. _P_ Straightening a paper clip (**Only the shape is changed. No new materials are formed.**)
9. _P_ Cloud formation (**Water vapor condenses to form liquid water — physical change only.**)
10. _C_ Rust on a piece of iron (**Oxygen combines with water and iron to form iron oxide—rust.**)
11. _C_ Separation of water into hydrogen and oxygen gases (**New materials are formed.**)
12. _P_ Dissolving sugar in water (**The water is still water and the sugar is still sugar, it is just broken into very small pieces and mixed up with the water.**)
13. _C_ Photosynthesis (**Water and carbon dioxide combine to form sugar and oxygen.**)
14. _C_ Bacteria decaying dead plant matter (**Complex tissues are broken down into basic elements by the bacteria.**)
15. _P_ Shine/luster (**Luster is determined by what the substance is. It is a physical characteristic.**)
16. _C_ A cake rising in the oven (**This is really both. Gas is produced as chemical reactions take place. There is also a physical element of the batter being stretched by the gas.**)
17. _P_ Cutting a piece of wood (**The wood is not changed into something else just because it becomes smaller.**)
18. _C_ Bread rising (**Like the cake, it is both chemical and physical.**)
19. _P_ Hardness (**This is a physical characteristic dependent on the material.**)
20. _P or C_ Making perfume (**Depends on the process.**)

11. States of Matter

What did we learn?

1. What are the three physical states of most matter? **Solid, liquid, gas.**
2. What is the name for each phase change? **Solid to liquid is melting, liquid to gas is evaporation, gas to liquid is condensation, liquid to solid is freezing, and for those substances that can go directly from solid to gas or gas to solid, the phase change is called sublimation.**
3. What is required to bring about a phase change in a substance? **The addition or removal of energy—primarily in the form of heat.**

Taking it further

1. Name several substances that are solid at room temperature. **The answers are endless. Some ideas include metals, wood, plastic, many foods, people, animals, etc.**

2. Name several substances that are liquid at room temperature. **Some ideas include water, juice, tea, honey, rubbing alcohol, and syrup.**

3. Name several substances that are gas at room temperature. **Some ideas include air, nitrogen, oxygen, hydrogen, carbon dioxide, carbon monoxide, propane, and natural gas.**

12. Solids

What did we learn?

1. What are three characteristics of solids? **They keep their shape, they have a definite volume, they are denser than most liquids and gases, they have lower kinetic energy than liquids or gases, and their molecules are closely packed together.**

2. How do large crystals form in solids? **If the liquid cools down very slowly, the molecules may be able to line up in regular patterns to make crystals.**

3. What state is the most common for the basic elements? **Nearly 90% of the elements are solids.**

Taking it further

1. Is gelatin a solid or a liquid? **Gels such as gelatin are neither a solid nor a liquid. Instead, they are a liquid that is suspended within a solid structure. Therefore, they have characteristics of both solids and liquids. They can be made to move like a liquid, yet hold their shape if left alone. Other phases can be suspended within each other as well. A foam is a gas that is suspended in a liquid. Smoke is a solid that is suspended in a gas. Fog is a liquid that is suspended in a gas.**

13. Liquids

What did we learn?

1. Which has more kinetic energy, a solid or a liquid? **A liquid.**

2. What shape does a liquid have? **The shape of its container.**

3. What is viscosity? **A measure of how strongly the liquid's molecules are attracted to each other.**

Taking it further

1. How is a liquid similar to a solid? **Both a solid and a liquid are much denser than a gas, both have a definite volume that can be measured, the molecules of both are close together.**

2. How is a liquid different from a solid? **Its molecules move freely over one another and its shape changes when you put it in a different container.**

3. How would you change a solid into a liquid? **You melt it by adding more energy, usually in the form of heat.**

14. Gases

What did we learn?

1. When is a substance called a gas? **When it has enough energy for the molecules to break apart from each other and move freely.**

2. What is the shape of a gas? **It takes on the shape of its container.**

3. In which state of matter are the molecules moving the fastest? **In a gas.**

4. What is atmospheric pressure? **The pressure applied to a surface by the collision of the air molecules with that surface.**

Taking it further

1. How is a gas similar to a liquid? **The molecules of both a gas and a liquid can move around, and they both take on the shape of their containers.**

2. How is a gas different from a liquid? **Gas molecules have much more energy, they freely move away from each other, and they collide with other molecules and objects billions of times a second. Gas expands to fill its container, so it does not have a definite volume.**

3. Why is it necessary that a spacesuit be pressurized in outer space? **God designed our bodies to operate in an environment where there is pressure on our bodies. If this pressure were not there, we would die. Since there is no air in space, there is no air pressure, so spacesuits must provide the pressure necessary for the astronauts.**

15. Gas Laws

What did we learn?

1. If temperature remains constant, what happens to the volume of a gas when the pressure is increased? **The volume decreases.**

2. If pressure remains constant, what happens to the volume of a gas when the temperature is increased? **The volume increases.**

3. What are two different ways to increase the volume of a gas? **Decrease the pressure or increase the temperature.**

Taking it further

1. Why might you need to check the air in your bike tires before you go for a ride on a cold day? **The volume of air may be decreased enough by the cold temperatures that you may need to add some air so your tires will not be flat.**

2. Why do you think increasing pressure decreases the volume of a gas? **The pressure forces the molecules closer together so they take up less space.**

3. Why do you think increasing temperature increases the volume of a gas? **The increase in the temperature adds energy to the molecules, causing them to move faster so they spread out more and take up more space.**

4. What might happen to the volume of a gas when the pressure is increased and the temperature is increased at the same time? **It depends on how much the pressure and temperature are increased. It is possible that the volume could remain the same. It could also increase or decrease. Because you are changing two things at once, you can't be certain of the effect without knowing how much you are changing each condition.**

16. Elements

What did we learn?

1. What is an element? **It is a substance that cannot be broken down by ordinary chemical means—an atom.**

2. What is a compound? **It is a substance that is formed when two or more elements combine chemically—a molecule.**

3. What is a mixture? **It is a combination of two or more substances that do not make a new substance.**

Taking it further

1. If a new element were discovered and named *newmaterialium*, would you expect it to be a metal or a nonmetal? **It would probably be a metal because most metal names end in "um" or "ium."**

2. Is salt an element, a compound, or a mixture? **Salt is a compound made from sodium and chlorine. It can be broken apart into its elements, but when they are put together, they form a new substance.**

3. Is a soft drink an element, a compound, or a mixture? **It is a mixture of water, sugar, flavorings, and other substances, but it is not a new substance.**

Challenge: Learning about Elements Worksheet

Name	Symbol	Atomic number
Hydrogen	H	1
Oxygen	O	8
Aluminum	Al	13
Silicon	Si	14
Mercury	Hg	80

Metal	Metalloid	Nonmetal
Sodium	Germanium	Nitrogen
Gold	Polonium	Phosphorus
Barium	Arsenic	Fluorine
Potassium	Antimony	Neon
Calcium	Boron	Chlorine
Silver		

Magnesium—12   Copper—29
Sulfur—16   Platinum—78
Argon—18   Bismuth—83
Iron—26   Radon—86

17. Compounds

What did we learn?

1. What is a compound? **A substance that is formed when two or more different kinds of atoms are chemically joined together.**

2. What is another name for an element? **An atom.**

3. What is another name for a compound? **A molecule.**

4. Do compounds behave the same way as the atoms that they are made from? **Not usually. Oxygen gas and hydrogen gas act very differently than liquid water or water vapor.**

Taking it further

1. The symbol for carbon dioxide is CO_2. What atoms combine to form this molecule? **One carbon atom and two oxygen atoms.**

2. The air consists of nitrogen and oxygen molecules. Is air a compound? Why or why not? **The air is not a compound because the nitrogen and oxygen molecules do not bond with each other to form a different substance. Instead, air is a mixture of gases.**

18. Water

What did we learn?

1. What two kinds of atoms combine to form water? **Hydrogen and oxygen.**

2. Why is water called a universal solvent? **Because a large variety of substances can be dissolved in water.**

3. What is unique about the water molecule that makes it able to dissolve so many substances? **The hydrogen atoms attach to the oxygen atom at a 105° angle, causing the charge to be unevenly distributed.**

Taking it further

1. What would happen to your body if oxygen could not be dissolved in water? **Your blood would not be able to take oxygen to the cells in your body, and you would die.**

2. Is water truly a universal solvent? **No, there are many substances, particularly oils and fats, that do not dissolve in water.**

3. Why is it important for mothers with nursing babies to drink lots of water? **Water is used in the production of milk.**

Water, Water Everywhere Worksheet

Food, water, beverages, washing dishes, cooking, making ice, growing house plants, water for pets, brushing teeth, toilets, bath/shower, washing hands/face, wiping counters, washing windows, laundry, mopping floors, watering grass, breathing, sweating/cooling your body, digestion, blood circulation, blinking, elimination of wastes, making of new cells, etc.

19. Mixtures

What did we learn?

1. What are two differences between a compound and a mixture? **A compound is formed when two or more elements combine to form a new substance. A mixture is formed when two or more elements or compounds are combined but do not form a new substance. The elements in a compound are always in the same proportion. The elements or compounds in a mixture can be in any proportion.**

2. What is a homogeneous mixture? **One in which all of the substances are evenly distributed.**

3. What is a heterogeneous mixture? **One in which all of the substances are not evenly distributed.**

4. Name three common mixtures. **Air, milk, granite, orange juice, seawater, etc.**

Taking it further

1. If a soft metal is combined with a gas to form a hard solid that doesn't look or act like either of the original substances, is the resulting substance a mixture or a compound? **The result is a compound because the new substance has different characteristics from the original substances. In a mixture, the substances retain their original properties.**

2. How might you separate the salt from the sand and water in a sample of seawater? **First, you could filter out the sand. Then you could let the water evaporate into the air and the salt would be left behind, or you could boil away the water like you did in lesson 10. This is similar to the experiment you did in lesson 6 with sugar and water.**

20. Air

What did we learn?

1. What is likely the most important element on earth? **Oxygen.**

2. What is likely the most important compound on earth? **Water.**

3. What is likely the most important mixture on earth? **Air.**

4. What are the main components of air? **Oxygen and nitrogen.**

Taking it further

1. Why is nitrogen necessary in the air? **Nitrogen dilutes the oxygen. Nitrogen also protects the earth from harmful gamma rays.**

2. Why is oxygen necessary in air? **Oxygen is necessary for cellular respiration in all plants and animals. Oxygen also protects the earth from harmful x-rays.**

3. How does the composition of air show God's provision for life? **Air provides exactly what is necessary for life without harming life on earth. Air protects living things from harmful radiation. Also, ozone, which is poisonous, is only found in the upper atmosphere.**

21. Milk & Cream

What did we learn?

1. Is milk an element, a compound, or a mixture? **Milk is a mixture.**

2. What is pasteurization, and why is it done to milk? **Pasteurization is the process of heating the milk to kill the bacteria in it.**

3. What is homogenization, and why is it done to milk? **Homogenization is the process that breaks the fat molecules into tiny bits so they stay suspended in the milk. This prevents the cream from separating from the milk and floating to the top.**

4. What is a foam? **It is a liquid that has air molecules suspended in it.**

Taking it further

1. Why does whipped cream begin to "weep"? **The fat molecules lose their ability to keep the lighter air molecules trapped, and the air eventually escapes.**

2. Why must cream be churned in order to make butter? **To form butter, the fat molecules must be forced together. Churning forces the molecules to clump together.**

22. Solutions

What did we learn?

1. What is a solution? **A mixture in which one substance is dissolved in another.**

2. Is a solution a homogeneous or heterogeneous mixture? **A solution is homogeneous.**

3. In a solution, what is the name for the substance being dissolved? **The solute.**

4. In a solution, what is the substance called in which the solute is dissolved? **The solvent.**

5. What is solubility? **The maximum amount of a substance that can be dissolved in a given amount of solvent.**

Taking it further

1. Why can more salt be dissolved in hot water than in cold water? **The warmer molecules are moving faster and can hold more salt molecules away from each other, so they can dissolve more salt than the slower, colder water molecules.**

2. If you want sweet iced tea, would it be better to add the sugar before or after you cool the tea? **If you add the sugar while the tea is hot, you will be able to dissolve more sugar and thus the tea will be sweeter. Whether this is better depends on how sweet you like your tea.**

Solutions Experiments Worksheet

Candy will dissolve fastest in hot water. Crushed candy will dissolve faster than whole candy. Candy will dissolve faster if you move your tongue because moving your tongue brings more saliva molecules in contact with the candy, so they have more opportunity to dissolve it.

Like Dissolves Like Worksheet

Soap is the only one of these substances that will dissolve in the oil because it is the only one that has a similar molecular structure to oil.

23. Suspensions

What did we learn?

1. What is a suspension? **A suspension is a mixture of substances that don't dissolve. It has particles of one substance that can stay suspended in the other for a short period of time, but not indefinitely.**

2. What does immiscible mean? **Two liquids that do not mix are immiscible.**

3. What is an emulsifier? **A substance that keeps immiscible liquids suspended.**

4. What is a colloid? **A liquid with very tiny particles suspended in it.**

Taking it further

1. What would happen to the mayonnaise if the egg yolk were left out of the recipe? **The oil would separate out, and it would lose its creamy texture.**

2. How is a suspension different from a true solution? **The molecules that are dissolved in a solution will stay dissolved indefinitely, whereas the particles that are suspended will eventually settle out of a suspension if an emulsifier is not added.**

24. Solubility

What did we learn?

1. What is solubility? **The ability of a solvent to dissolve a solute.**

2. What are the three factors that most affect solubility? **The type of materials being dissolved, temperature, and pressure.**

3. What is the name given to particles that come out of a saturated solution? **A precipitate or precipitation.**

Taking it further

1. Why are soft drinks canned or bottled at low temperatures and high pressure? **Soft drinks are a solution of carbon dioxide dissolved in a liquid. To keep the maximum amount of gas dissolved, the solution is canned or bottled at low temperatures under high pressure.**

2. Why do soft drinks eventually go flat once opened? **The pressure has been reduced on the solution, so the liquid cannot hold as much gas as it once did. The gas escapes into the air, and the drink tastes flat.**

3. If no additional sugar has been added to a saturated solution of sugar water, what can you conclude about the temperature and/or pressure if you notice sugar beginning to settle on the bottom of the cup? **You can conclude that the temperature of the solution has dropped.**

Solubility of Various Substances Worksheet

Sugar has the highest solubility; baking soda has the lowest solubility.

25. Soft Drinks

What did we learn?

1. What are the main ingredients of soft drinks? **Water, sweetener, flavoring, color, and carbon dioxide.**

2. What is the most popular drink in the United States? The second most popular? **Water, followed in second place by soft drinks.**

3. What are the two most popular sweeteners used in soft drinks? **Corn syrup and aspartame.**

Taking it further

1. Why are soft drink cans warmed and dried before they are boxed? **The drink is very cold when it is canned or bottled. As it warms up, water condenses on the outside of the can or bottle. If this occurred after packaging, the water would make the boxes or cartons soggy, so it is done beforehand.**

2. Why are recipes for soft drinks considered top secret? **People buy a particular brand of soft drink because they like that flavor better than any other. So if someone obtained a secret recipe, the original company could lose money.**

3. Why would the finished syrup be tested before adding the carbonation? **To ensure that it tastes correctly before it completes the process; to make sure that nothing went wrong in the previous steps.**

26. Concentration

What did we learn?

1. What is a dilute solution? **One in which there are relatively few solute molecules in the solution.**

2. What is a concentrated solution? **One in which there are a relatively large number of solute molecules in the solution.**

3. How does the concentration of a solution affect its boiling point? **In general, the more concentrated it is, the higher the boiling point will be.**

4. How does the concentration of a solution affect its freezing point? **In general, the more concentrated it is, the lower the freezing point will be.**

Taking it further

1. Why is a quantitative observation for concentration usually more useful than a qualitative observation? **Qualitative observations are based on people's perceptions and not easily repeated. One person may think that the lemonade is too strong while another thinks it is too weak. But quantitative observations are not a matter of opinion and can be repeated.**

2. If a little antifreeze helps an engine run better, would it be better to add straight antifreeze to the radiator? Why or why not? **Not necessarily. The combination of different molecules raises the boiling point and lowers the freezing point of both substances in the solution, but straight antifreeze would not necessarily have the same effect.**

Challenge: Salt's Effect on the Freezing and Boiling Point of Water Worksheet

You should find that salt increases the boiling point and decreases the freezing point of water.

27. Seawater

What did we learn?

1. What is the most common solution on earth? **Seawater.**

2. What are the main substances found in the ocean besides water? **Sodium chloride—salt, magnesium, and bromine.**

3. How does salt get into the ocean? **Water flowing over land dissolves salt and other minerals and carries them to the ocean. The minerals stay behind when the water evaporates.**

4. Name one gas that is dissolved in the ocean water. **Oxygen is the main gas. Nitrogen, carbon dioxide, and other gases are present as well.**

Taking it further

1. Why is seawater saltier than water in rivers and lakes? **Fresh water is continually entering and exiting the rivers and lakes, so the amount of salt remains low. However, in the ocean, the only way that water leaves is through evaporation, which removes the water but leaves the minerals. After thousands of years, the salt has built up in the oceans.**
2. Why is there more oxygen near the surface of the ocean than in deeper parts? **Algae and other plants grow near the surface and produce oxygen that dissolves in the water.**

28. Water Treatment

What did we learn?

1. Why do we need water treatment plants? **Water from rivers and lakes contains dirt, harmful bacteria, and other substances that are not healthy for people to drink.**
2. What are the three main things that are done to water to make it clean enough for human consumption? **Particles are allowed to settle out, chemicals are added to kill bacteria, and the water is filtered.**
3. Why is it important not to dump harmful chemicals into rivers and lakes? **The chemicals will dissolve in the water and harm the plants and animals living there.**

Taking it further

1. How is the filter you built similar to God's design for cleaning the water? **Much of the water that falls on the earth sinks into the ground, where it flows through sand and gravel and becomes cleaner before reaching rivers and underground water tables.**

29. Food Chemistry

What did we learn?

1. What are the three main types of chemicals that naturally occur in food? **Carbohydrates, proteins, and fats.**
2. What kinds of chemicals are often added to foods? **Preservatives, flavor enhancers, and color enhancers.**
3. Why is the kitchen a great place to look for chemicals? **All of our foods are made of chemicals, and many chemical reactions occur as we are cooking.**

Taking it further

1. If you eat a peanut butter and jelly sandwich, which part of the sandwich will be providing the most carbohydrates? The most fat? The most protein? **The bread will provide the most carbohydrates, although it depends on how much jelly you put on. Jelly is mostly sugar, which is also a carbohydrate. The peanut butter will provide nearly all of the fat and most of the protein.**

Challenge: Food Chemicals Worksheet

1. What chemical is found in coffee and soft drinks that interferes with some people falling asleep? **Caffeine.**
2. What is the chemical name for table sugar? **Sucrose.**
3. What is the chemical name for baking soda? **Sodium bicarbonate.**
4. What is the chemical name for table salt? **Sodium chloride.**
5. What chemical makes you cry when you slice onions? **Sulfuric acid.**
6. What chemical gives peppers their hot flavor? **Capsaicin.**
7. What chemical gives carrots their orange color? **Carotene.**
8. What chemical gives tomatoes their red color? **Lycopene.**
9. What chemical gives broccoli its green color? **Chlorophyll.**
10. What chemical gives a soft drink its bubbles? **Carbon dioxide.**

30. Chemical Analysis of Food

What did we learn?

1. What are the main chemicals listed on food labels? **Carbohydrates such as sugar and starch, proteins, fats, vitamins, and minerals.**
2. How do food manufacturers know what to put on their labels? **Chemists have tested the foods to see what they are composed of.**

3. What is one way to test if a food contains oil? **Place a sample on brown paper for a few minutes and see if it makes the paper become translucent.**

4. What is an indicator? **A substance that is used to detect the presence of a particular chemical, usually by changing color.**

Taking it further

1. How do you suppose indicators work? **Usually the indicator molecules react with the desired chemical to produce a substance that is a different color from the indicator. For example, iodine turns blue in the presence of starch molecules because of a chemical reaction with the starch that produces a blue substance.**

2. Why is it important to know what chemicals are in our food? **This information allows us to compare different foods and decide which ones are best to eat. Also, some people are allergic to particular foods, and food labels help them avoid those foods.**

Chemical Analysis Worksheets

- Vegetable oil, peanut butter, and chips should all have turned the paper translucent since these foods are high in fat/oil. Depending on the type of bread you used, you may have detected a small amount of oil.

- Bread, corn, potatoes, and flour are all high in starch and should have changed color when combined with iodine.

- Water does not contain any fats or starches, so it should not have reacted in either test. Apples contain a high amount of sugar but very little fat and starch, so they should not have reacted in either test.

31. Flavors

What did we learn?

1. What two parts of your body are needed in order to fully enjoy the flavor of your food? **The taste buds in your mouth and the smell receptors in your nose.**

2. What is the difference between an herb and a spice? **Herbs come from the leaves of a plant; spices come from other parts of the plant.**

3. What is the difference between a natural flavor and an artificial flavor? **Natural flavors come directly from a plant. Artificial flavors are created by combining chemicals in a lab or factory.**

Taking it further

1. Why might a cook prefer to use fresh herbs rather than dried herbs? **Fresh herbs usually have a milder flavor than dried herbs. However, fresh herbs can spoil more quickly than dried herbs.**

2. Why do you think artificial vanilla tastes different than natural vanilla even though they may have the same chemical formula? **Flavor is a complicated thing. Scientists are not quite sure how the flavors are changed, but artificial flavors often produce an unpleasant or bitter aftertaste that natural flavors do not have.**

32. Additives

What did we learn?

1. What is a food additive? **Anything that is added to the food by a manufacturer.**

2. Name three different kinds of additives. **Preservatives, antioxidants, emulsifiers, stabilizers, coloring, flavor enhancers, vitamins, minerals, etc.**

3. Why are preservatives sometimes added to foods? **To keep the foods from spoiling.**

4. What compound has been used as a preservative for thousands of years? **Salt has been used to preserve many foods, especially meats. Sugar has also been used for a long time.**

5. Why are emulsifiers sometimes added to foods? **To keep the oil and water in the foods from separating. Remember when you made mayonnaise?**

Taking it further

1. Why are vitamins and minerals added to foods? **Processing, such as heating, often kills bacteria but also destroys many of the nutrients in the foods. Vitamins and minerals are often added back in to restore the nutritional value of the food.**

2. Why does homemade bread spoil faster than store-bought bread? **Because it does not contain preservatives like the store-bought bread does.**

33. Bread

What did we learn?

1. If you want fluffy bread, what are the two most important ingredients? **Wheat flour that contains gluten, and yeast.**

2. Why is gluten important for fluffy bread? **The gluten allows the bread dough to stretch and traps the gas produced by the yeast.**

3. Why does bread have to be baked before you eat it? **The baking process breaks down the long starch molecules into smaller molecules that are more easily digested.**

4. Why is whole wheat bread more nutritious than white bread? **The white flour does not contain all of the parts of the wheat kernel, so it has fewer nutrients.**

Taking it further

1. What would happen if you did not put any sugar in your bread dough? **The yeast would not be able to produce as much carbon dioxide gas, so your bread would not be as fluffy.**

2. Can bread be made without yeast? **Yes, other forms of leavening can be used such as baking soda or baking powder. However, the bread will be more like tortillas or pita bread than the fluffy bread you may be used to.**

Challenge: Homemade vs. Store-bought Worksheet

Homemade bread, in general, will dry out faster and will grow more mold than store-bought bread. However, you may prefer the flavor of homemade bread. Also, homemade bread may be more nutritious.

34. Identification of Unknown Substances: Final Project

What did we learn?

1. What method should be used in identifying unknown substances? **The scientific method.**

2. Why should you avoid tasting unknown substances? **The substance can be dangerous or harmful, so you don't want to taste it if you don't know what it is.**

3. How can you test the scent of an unknown substance safely? **Hold it a few inches away from your nose and push some air toward your nose. This allows you to smell a few molecules without damaging your nose if the scent is very strong or caustic, like ammonia.**

4. What are some physical characteristics of an unknown substance you can test at home? **Mass, density, melting point, freezing point, boiling point, and state—such as solid, liquid, or gas.**

5. What are some chemical characteristics you can test at home? **Presence of starch, oil, baking powder, acid, or base.**

Taking it further

1. Why is it important for food manufacturers to test the ingredients they use and final products they produce? **To ensure the safety and flavor of their foods.**

2. Why is it important for water treatment facilities to test the quality of the water? **We don't want harmful bacteria or other dangerous substances in our water supply.**

35. Conclusion

What did we learn?

1. What is the best thing you learned about matter? **Answers will vary.**

Taking it further

1. What else would you like to know about matter? **Go to the library and learn about it.**

Properties of Ecosystems → Worksheet Answer Keys

1. What Is an Ecosystem?

What did we learn?

1. What is ecology? **The study of plants and animals and the environment in which they live.**
2. What is the biosphere? **The part of the earth in which living things exist—includes the atmosphere, surface of the earth, underground, and the water.**
3. Give an example of something that is biotic and something that is abiotic. **Examples of biotic: plants, animals, fungi, bacteria. Examples of abiotic: rocks, man-made objects, soil, weather.**
4. What is flora? **Plants.**
5. What is fauna? **Animals.**

Taking it further

1. What factor has the greatest effect on the plants and animals that live in a particular ecosystem? **The climate.**
2. How does your habitat change throughout the day? **Moving from the home to school or a store, going to the park or other area of activity.**
3. List some ways that climate affects the habitats of people. **The houses they live in, the clothes they wear, the activities they participate in, and the foods that are readily available.**

2. Niches

What did we learn?

1. What is a niche? **The roles played by the plant or animal within its environment.**
2. Name two factors that determine an animal's niche. **What it eats, what eats it, how it acts, things it can do, and its relationships with other animals.**
3. What is a population? **The total number of a single species in a given area.**
4. What is a community? **All of the populations in a given area.**
5. What are two different kinds of niches an animal can have? **The niche a species has within the whole community and the niche a particular organism has within its species/colony.**

Taking it further

1. What different niches do you fill in your family and in your community? **Child, sibling, cook, student, team member, performer, etc.**
2. How does competition for food and other resources affect the niche of a plant or animal? **Competition occurs when there are limited resources. This limits the population of a species. It may result in certain plants or animals being aggressive or having specialized roles.**

3. Food Chains

What did we learn?

1. What is a food chain? **A series of organisms in the order in which they feed on one another.**
2. What is a producer? **A plant—something that makes its own food.**
3. What is a consumer? **An organism that feeds on other organisms.**
4. What is a food web? **Interconnecting food chains, etc.**
5. List two herbivores. **Deer, antelope, cattle, horses, etc.**
6. List two carnivores. **Wolf, coyote, weasel, lion, snake, etc.**
7. List two omnivores. **Bear, man, raccoon, mice, etc.**

Taking it further

1. Is a black bear a first or second order consumer? **It depends on what it is eating. If it is eating plants, it is a first order consumer. If it is eating fish, it is a second or maybe even third order consumer, depending on what the fish ate.**
2. Is man an herbivore, carnivore, or omnivore? **Some people choose to live a vegetarian lifestyle, so they would be considered herbivores, but most people eat producers and consumers and would be considered omnivores.**
3. Explain how a food chain shows energy flow. **A food chain starts with a plant, which converts sunlight into energy. That energy is passed on to the animal that eats the plant. Some of that energy is used up and some becomes part of the animal's body. That energy is then passed on to the next consumer.**

4. Scavengers & Decomposers

What did we learn?
1. What are organisms called that eat dead plants and animals? **Scavengers.**
2. Name two different animals that eat dead plants or animals. **Vultures, flies, earthworms, coyotes, opossums, etc.**
3. What types of organisms are at the end of every food chain? **Decomposers.**
4. Name two common organisms responsible for decomposition. **Bacteria and fungi.**

Taking it further
1. Why is decomposition so important? **It is the process that frees up the elements that were stored in the tissues of the dead plant or animal so they can be recycled.**
2. What physical law makes decomposition necessary? **The law of conservation of matter/mass.**

5. Relationships among Living Things

What did we learn?
1. What is symbiosis? **A close relationship between two different species.**
2. What is mutualism? **A symbiotic relationship in which both species benefit from each other.**
3. What happens to each species in a parasitic relationship? **The guest benefits and the host is harmed.**
4. Which species benefits in commensalism? **The guest species benefits.**
5. What is competition among species? **When two species compete for limited resources.**
6. What is the name of a relationship in which neither species benefits nor is harmed? **Neutralism.**

Taking it further
1. Why is competition considered harmful for both species? **When their resources are limited, some plants or animals will not get what they need and may die or fail to reproduce. This could affect both species that are competing for the resources.**
2. Explain how competition could keep the species from becoming too populated. **When there are not enough resources for everyone to live, some plants or animals will die or fail to reproduce. This will prevent the population from becoming too large.**

Symbiosis Worksheet

		Species B		
Species A		+	0	-
	+	Mutualism	Commensalism	Parasitism
	0	Commensalism	Neutralism	XXX
	-	Parasitism	XXX	Competition

6. Oxygen & Water Cycles

What did we learn?
1. How do photosynthesis and respiration demonstrate the oxygen cycle? **During photosynthesis, carbon dioxide and water, which contain oxygen atoms, are absorbed. Glucose and oxygen are produced. Animals eat the glucose and breathe in the oxygen. During respiration, these molecules are broken down to release the energy and to produce water and carbon dioxide for plants to use again.**
2. What are the major steps in the water cycle? **Evaporation, condensation, and precipitation.**

Taking it further
1. Water exists in three forms: solid, liquid, and gas. What phase is the water in before and after evaporation? **It changes from liquid into gas.**
2. What phase is the water in before and after condensation? **It changes from gas to liquid.**
3. What phase is the water in before and after precipitation? **If the temperature is not too cold, it stays as a liquid. If the temperature is cold enough, the water can change from liquid to solid and comes down as snow or sleet.**

7. Biomes around the World

What did we learn?
1. Where is the tropical zone located? **Between the Tropic of Cancer (23.5° north latitude) and the Tropic of Capricorn (23.5° south latitude), centered on the equator.**
2. Where is the northern temperate zone located? **Between the Tropic of Cancer and the Arctic Circle (66.5° N).**

3. Where is the southern temperate zone located? **Between the Tropic of Capricorn and the Antarctic Circle (66.5° S).**

4. Where are the polar regions located? **North of the Arctic Circle and south of the Antarctic Circle.**

Taking it further

1. Why are the polar regions generally colder than the tropical regions even though they receive many more hours of sunlight each day during the summer? **The sunlight reaches the earth at a sharper angle and much of it reflects away from the surface of the earth. Also, snow and ice tend to reflect much of the sunlight rather than absorbing it, thus keeping the temperature colder.**

2. What correlations do you see between the temperature and rainfall maps that you made? **Answers will vary, but there is generally more rainfall in warmer areas, excepting deserts.**

8. Grasslands

What did we learn?

1. Name three characteristics of a grassland biome. **10–30 inches of rain per year, distinct wet and dry seasons, warm summers and cold winters, and grass is the primary plant with few trees and shrubs.**

2. What are four different types of grasslands? **Prairie, savannah, pampas, and steppe.**

3. Where can each of these grasslands be found? **Prairie—North America; Savannah—Africa; Pampas—South America; Steppe—Europe and Asia.**

Taking it further

1. Why are there few trees in a grassland? **There is not enough rain to support trees. Also, periodic fires kill trees and shrubs.**

2. How do many plants survive extended periods of drought in the grassland? **Many plants, such as grass, become dormant until there is enough water to resume growth. Other plants have very long roots to reach water deep underground.**

3. How can grass survive when it is continually being cut down by grazing animals? **The growth center of grass is at the bottom of the plant, near ground level, so it can continue to grow after its top is cut off.**

Challenge: Growing Grass Worksheet

- How did cutting the grass affect its ability to grow? **It didn't.**
- Did one plant grow more than the others? **Answers will vary.**
- How does this experiment demonstrate God's provision for grassland animals? **Grass provides food for grazing animals and continues to provide more food even when eaten over and over again.**

9. Forests

What did we learn?

1. What are the major plants in a forest? **Trees.**

2. What are the six layers of a forest? **Emergent layer, canopy, understory, shrub, herb, and floor.**

3. Which layer forms the roof of the forest? **Canopy.**

4. Name three kinds of forests. **Deciduous, coniferous, tropical rainforest.**

Taking it further

1. Why is the forest floor relatively dark? **The trees grow close enough together for their leaves to block out much of the light.**

2. Why is it important to study each layer of a forest? **Different plants and animals can be found in each layer, so you must study all the layers in order to understand the whole ecosystem.**

3. How might new trees find room to grow in a mature forest? **Room is made when older trees die and fall down or when trees are damaged in a storm. Also, trees can be cut down by people.**

Where Would I Live? Worksheet

Emergent Layer	Bald eagle, flies
Canopy	Spider monkey, opossum, fruit bat, tree frog, lemur, flies, woodpecker
Understory	Fruit bat, hummingbird, tree frog, flies, woodpecker
Shrub Layer	Monarch butterfly, hummingbird, flies
Herb Layer	Monarch butterfly, hummingbird, flies
Floor	Termites, black bear, rabbit, ibis, deer, flies

10. Temperate Forests

What did we learn?

1. What are some characteristics of a deciduous forest? **Dominant plants are deciduous trees, 30–60 inches of rainfall per year, four distinct seasons, warm wet summer and cold winter, trees lose leaves in the fall.**

2. What are some characteristics of a coniferous forest? **Dominant plants are evergreen/coniferous trees, 12–33 inches of rainfall/precipitation per year, cold winters, many lakes.**

3. What is another name for a coniferous forest in the far north? **Boreal forest or taiga.**

4. What is a deciduous tree? **One that has broad, flat leaves and sheds its leaves in the fall.**

5. What is a coniferous tree? **One that has needle-like leaves that do not fall off and has cones instead of flowers.**

Taking it further

1. What are some ways that plants in temperate forests were designed to withstand the cold winters? **Deciduous trees lose their leaves; coniferous trees have needles that are not damaged by freezing temperatures. Trees have thick bark that helps protect them from cold wind and snow.**

2. What are some ways that animals in temperate forests were designed to withstand the cold winters? **Many animals hibernate; others go into a deep sleep. Still others migrate to warmer areas during the winter and return in the summer months.**

3. Would you expect plant material that falls to the floor of the coniferous forest to decay quickly or slowly? Why? **The floor of the coniferous forest is relatively dry and often cold. This is not an ideal environment for bacteria to grow, so material decays relatively slowly in the coniferous forest.**

Challenge: Forest Jeopardy Worksheet

Accept all reasonable questions.

1. Oak, maple, and beech. **What kinds of trees might you find in a deciduous forest?**

2. Roof of the forest. **What is another name for the canopy of the forest?**

3. Lichen, moss, and fungi. **What plants might you find growing on the floor of a forest?**

4. Shrub layer. **What layer in the forest is below the understory?**

5. 30–60 inches per year. **How much rain does a deciduous forest receive each year?**

6. 12–33 inches per year. **How much rain/precipitation does a coniferous forest receive each year?**

7. Tropical and polar regions. **Between which two regions do you find temperate forests?**

8. Boreal forest and taiga. **What other names refer to a coniferous forest?**

9. Dall and big horn sheep. **What are some animals that are likely to live in a coniferous forest?**

10. Many lakes. **What geologic features are found in coniferous forests as a result of the glacier movement during the Ice Age?**

11. Duck-billed platypus. **What is one animal that is found only in the forests of Australia?**

12. Tallest trees of the forest. **What would you see in the emergent layer of the forest?**

11. Tropical Rainforests

What did we learn?

1. List some ways in which a tropical rainforest is different from a temperate forest. **Rainforest receives more rain (over 80 inches per year). Rainforest is always warm to hot—no cold winters. Rainforests have more different kinds of animals, but relatively fewer mammals.**

2. Where are the rainforests located? **Between the Tropic of Cancer and the Tropic of Capricorn; in the tropical region.**

3. What is an arboreal animal? **One that lives primarily in trees.**

4. What is an epiphyte? **A plant that grows on another plant without taking nutrients from it.**

5. Name at least one epiphyte. **Orchid, fern, cactus, banyan tree.**

Taking it further

1. Do you think that dead materials would decay slowly or quickly on the floor of the rainforest? Why? **Because the rainforest is warm and moist all the time, bacteria and other composters thrive; thus, dead material decays very quickly.**

2. If you transplanted trees such as orange, cacao, or papaya trees to a deciduous forest, would you expect them to survive? Why or why not? **Many**

tropical plants cannot survive the cold winters that are experienced in the deciduous forests. These plants would not be likely to survive.

3. Which animals are you most likely to see if you are taking a walk through the tropical rainforest? **Because many animals live primarily in the canopy, you would only see the ones that live near the floor or that visit the floor regularly. These might include lizards and snakes, capybaras, a few birds, and lots of insects. You would probably not see monkeys except from afar.**

12. The Ocean

What did we learn?

1. How much of the earth is covered with water? **About 75%.**
2. How much of the surface water of the world is in the ocean? **About 97%.**
3. How many oceans are there? **Although there are five named oceans, they are all connected, making only one ocean.**
4. What are the three zones that the ocean can be divided into? **Sunlit (euphotic), twilight (disphotic), and midnight (aphotic) zones.**
5. What are the three major groups of living organisms in the ocean? **Benthos, nekton, and plankton.**

Taking it further

1. What might happen in the ocean if the currents stopped flowing? **The plankton would not be moved around and some areas of the ocean would have less food than now. This would cause some animals to die or leave the area. Nutrients in one area would be used up and plankton would die, causing other animals to die.**
2. Why do most animals in the ocean live in the euphotic zone? **Photosynthesis can only take place where there is sufficient sunlight, so food is most abundant in the euphotic zone. Therefore, most animals will be found there.**
3. Why might the aphotic zone occur at a shallower depth than 660 feet (200 m) in some areas? **The amount of sunlight that can penetrate the water depends on how clear the water is. If there is a significant amount of silt or other particles in the water, this will reduce the depth that the sunlight can penetrate.**

13. Coral Reefs

What did we learn?

1. Where will you find coral reefs? **In warm, clear water near the equator.**
2. What is a coral reef made from? **Limestone from the exoskeletons of coral.**
3. Where do corals get most of their energy? **From the algae that live in them.**
4. What are the three main types of coral reefs? **Atoll, fringing, and barrier.**
5. What are some of the animals that live in a coral reef besides corals? **Sponges, shrimp, sea stars, eels, turtles, octopus, fish, whales, etc.**

Taking it further

1. Why are coral reefs found in water that is usually less than 150 feet (45 m) deep? **The algae in the coral require sunlight for photosynthesis, so coral cannot survive where there is not enough sunlight.**
2. Why do corals grow best in swift water? **The moving water brings more nutrients, which spurs growth.**

14. Beaches

What did we learn?

1. What is a beach? **The area where the water meets the land.**
2. What are the two main kinds of beaches? **Rocky and sandy.**
3. What is the name of the area of land that is covered at high tide and uncovered at low tide? **The intertidal zone.**
4. What are some animals you are likely to see in a beach ecosystem? **Clams, mussels, crabs, oysters, starfish, barnacles, turtles, and gulls.**

Taking it further

1. Why might you find different plants and animals on a rocky beach from those on a sandy beach? **A rocky beach provides more places for plants to anchor, so a wider variety of plants and animals is likely to survive there.**
2. How is new sand formed? **Waves erode rocks, shells, and coral to form new sand. Also, new sand can be formed when hot lava flows into cold water.**

3. Explain how a beach can be in dynamic equilibrium. **Sand is made and deposited by the action of the waves while, at the same time, other sand and materials are dragged out to sea by the tide. If the amount deposited is about equal to the amount removed, the beach is said to be in dynamic equilibrium.**

15. Estuaries

What did we learn?

1. What is an estuary? **An area where fresh water flows into saltwater.**

2. Name three types of estuaries. **Salt marsh, salt meadow, and mangrove forest.**

3. What are some plants you might find in an estuary? **Reeds, salt grass, mangrove trees, etc.**

4. Name several animals that you might find in an estuary. **Mud snails, marine worms, shellfish, mullet, flounder, sole, herons, terns, storks, pelicans, sea lions, etc.**

Taking it further

1. Why is an estuary a very productive ecosystem? **The moving water stirs up nutrients that spur plant growth.**

2. How do mangrove trees help coral reefs? **The trees help to filter out silt that might otherwise make the water cloudy.**

3. Why is the salt level in the water constantly changing in an estuary? **Fresh water and saltwater do not easily mix. There is a constant flow of fresh water and a changing flow of saltwater due to tides, so the salt level is changing. Seasonal changes in weather also affect salt levels.**

4. Why might you find different estuary animals in the same location at different times of the year? **Many animals migrate and spend different parts of the year in different locations.**

16. Lakes & Ponds

What did we learn?

1. What is a lake? **A large body of fresh water.**

2. What is a pond? **A lake that is not deep enough to have a dark zone.**

3. What are two ways that lakes were formed in the past? **Some lakes were dug out by glaciers; others have formed in craters of extinct volcanoes. It is likely that many of the lakes formed as a result of the Great Flood.**

4. What is an overturn? **It is when colder water on the top of a lake rapidly sinks, causing warmer water to rise.**

5. What is an algae bloom? **A rapid growth in algae.**

Taking it further

1. Why is overturn important to lake ecosystems? **It releases nutrients and oxygen that become trapped in the mud at the bottom of the lake.**

2. Why does an algae bloom often occur in a lake in the spring? **Overturn occurs in the spring and releases nutrients that algae need to grow, causing algae to grow quickly.**

3. In which lake zone would you expect to find most small creatures like rotifers? **They will most likely be in the sunlit zone because they eat algae, and algae need sunlight.**

4. What would happen to fish during the winter if ice did not float? **As ice began to fill up the bottom of the lake, the fish would be forced to move up in the lake. Eventually the whole lake could freeze, and the fish would die.**

Challenge: Great Lakes Fact Sheet

Answers may vary depending on the source. These numbers are from the EPA.

Feature	Lake Superior	Lake Michigan	Lake Huron	Lake Erie	Lake Ontario
Average Depth	483 feet 147 meters	279 feet 85 meters	195 feet 59 meters	62 feet 19 meters	283 feet 86 meters
Maximum Depth	1322 feet 406 meters	925 feet 282 meters	750 feet 229 meters	210 feet 64 meters	802 feet 244 meters
Volume	2900 miles^3 12,100 km^3	1180 miles^3 4920 km^3	850 miles^3 3540 km^3	116 miles^3 484 km^3	393 miles^3 1640 km^3
Major cities that border it	Duluth, MN Sault Ste. Marie, ON Thunder Bay, ON Marquette, MI	Chicago, IL Gary, IN Green Bay, WI Milwaukee, WI	Sarnia, ON Port Huron, MI Bay City, MI	Buffalo, NY Cleveland, OH Érie, PA Toledo, OH	Hamilton, ON Kingston, ON Oshawa, ON Rochester, NY Toronto, ON Mississauga, ON

17. Rivers & Streams

What did we learn?

1. What is a river? **A moving body of fresh water.**

2. Where does most of the energy for a river ecosystem come from? **Plant material that falls into the river.**

3. Name some plants you might find in a river ecosystem. **Grasses, pussy willows, alders, elkslip, willow trees, etc.**

4. What is a tributary? **A smaller river or stream that flows into a larger river.**

5. What is the riparian zone? **The area along the banks of a river.**

Taking it further

1. Why do fewer plants grow in the water of a river than in a lake or ocean? **The current of the river makes it difficult for plants to stay in one place, and most plants need to stay anchored to survive.**

2. Would you expect a river to be larger at a higher elevation or a lower elevation? **In general, because tributaries are adding water to a river as it flows downhill, you would expect the river to be smaller at higher elevations and larger at lower elevations.**

3. Do rivers move faster over steep ground or in relatively flat areas? **Gravity is what causes water to flow, so water will flow faster over steeper ground.**

4. Would you expect water to cause more erosion in a steep area or in a relatively flat area? **The faster water is moving, the more erosion it can cause, so more erosion will occur in steeper areas.**

Challenge: River Facts Sheet

See page 432.

18. Tundra

What did we learn?

1. Where is most tundra located? **In the northern regions of Alaska, Canada, Greenland, Scandinavia, and Russia.**

2. What is permafrost? **The layer below the surface that never thaws, even in summer.**

3. What kind of plants grow in the tundra? **Small plants, including flowers, small shrubs, rushes, sedges, heather, mosses, and lichens.**

4. What are some animals you might find in the tundra? **Polar bears, Arctic foxes, caribou, moose, ptarmigan, Canada geese, Arctic hares, flies, mosquitoes, etc.**

5. How much precipitation does the tundra receive? **6–10 inches a year.**

Taking it further

1. Why do many animals in the tundra have white fur or feathers? **To provide them with camouflage from predators in the snow.**

2. Why do many animals and plants have an accelerated life cycle in the tundra? **Because the growing season is only 50–60 days long.**

3. Why do you think the temperatures are so cool in the summer when there is often 24 hours of sunshine? **Although the sun is up for many weeks, its light reaches the earth at a steep angle in the tundra, so the energy is spread out. Also, the ice and snow reflect much of the light away from the ground, thus keeping it cool.**

19. Deserts

What did we learn?

1. What is a desert ecosystem? **One which receives less than 10 inches of rain per year.**

2. How is a cold desert different from a hot desert? **Daytime temperatures drop below freezing in the winter in a cold desert but remain significantly above freezing during the day in a hot desert.**

3. What are some plants you would expect to find in the desert? **Cactus, sagebrush, aloe, mesquite, Joshua tree, creosote bush, desert trumpet, etc.**

4. What are some animals you would expect to find in the desert? **Mouse, toad, snake, lizard, badgers, ostriches, vultures, owls, coyotes, etc.**

5. What is the difference between a Bactrian camel and a dromedary camel? **Bactrian camels have two humps and longer hair; dromedary camels have short hair and one hump.**

Taking it further

1. In what ways are plants well suited for the desert environment? **Some can store large amounts of water; some have needles that do not lose water through transpiration; others have leaves with very few stomata; many have accelerated life cycles.**

2. In what ways are animals well suited for the desert environment? **Most are nocturnal; some estivate; many have an accelerated life cycle.**

3. Why does rain often cause flash flooding in the desert? **The ground is so dry and hard that water does not quickly soak in.**

4. What are some dangers you may face in the desert? **Dehydration due to lack of water, heat stroke, freezing/exposure due to cold temperatures when the sun goes down, sand storms, and scorpion stings.**

5. Why do salt flats often form in the desert? **Since water does not quickly soak into the ground, much of it evaporates, leaving dissolved salt behind. Over hundreds of years, this salt builds up to form salt flats.**

6. Would you expect to find more salt flats in a cold desert or a hot desert? **More salt flats are found in cold deserts because cold deserts usually receive more water than hot deserts.**

20. Oases

What did we learn?

1. What is an oasis? **An ecosystem in the desert where water is readily available.**

2. What kinds of plants grow in an oasis? **Palm trees, shrubs, grass, and cacti.**

3. What kinds of animals live in an oasis that don't usually live in a desert? **Fish, bats, warblers, and orioles.**

Taking it further

1. Why is it often cooler in an oasis than in a desert? **The transpiration from the trees results in evaporation, which cools the air. Also, the leaves of the trees block some of the sun.**

2. Why are oases important for trade routes? **The only way to safely cross the desert in the past was by traveling from one oasis to another.**

3. How might a man-made oasis change the ecosystem in a desert? **The water that is brought in will make it possible to grow plants that do not naturally grow there. This will provide habitat for animals that do not naturally live there. Also, it will add humidity to the air, thus cooling it down and possibly increasing the rainfall.**

21. Mountains

What did we learn?

1. What ecosystems are you likely to encounter on mountains in temperate zones? **Grasslands, deciduous forests, evergreen forests, alpine meadows, and alpine tundra.**

2. What ecosystems are you likely to encounter on mountains in tropical zones? **Rainforests, bamboo forests, heath, meadows, and tundra.**

3. What is timberline? **The point above which no trees will grow.**

4. What is snow line? **The point above which the snow does not completely melt, even in the summertime.**

Taking it further

1. Why do the ecosystems change as you gain altitude on a mountain? **Temperature, rainfall, and oxygen levels change as you gain altitude, so different plants and animals will live at different altitudes.**

2. Why don't you find every ecosystem on every mountain? **Different mountains are different heights. Most mountains are not high enough to experience all of the different ecosystems.**

3. What other ecosystems are you likely to find on mountains that were not listed in this lesson? **Rivers, lakes, and ponds are abundant in most mountains.**

4. How have glaciers influenced the shapes of mountains? **As glaciers receded at the end of the Great Ice Age, they dug out valleys, lakes, and other features in the mountains.**

5. Why is there less oxygen as you gain altitude? **The gravitational pull of the earth becomes less as you go away from the center of the earth, so fewer air molecules are held close to the earth at higher altitudes.**

22. Chaparral

What did we learn?

1. What is a chaparral ecosystem? **An ecosystem on hot, dry slopes in areas with mild, rainy winters.**

2. What are two other names for a chaparral? **Mediterranean ecosystem or maquis.**

3. Name some plants you might find in the chaparral. **Scrub oak, live oak, yucca, buckbrush, trefoil, etc.**

4. Name some animals you might find in the chaparral. **Woodrat, rabbit, fox, coyote, bobcat, quail, jay, wren, sparrow, etc.**

5. What animal might you find in the Australian chaparral that you would not find in the American chaparral? **Koala.**

Taking it further

1. What conditions make fire likely in the chaparral? **Hot, dry summers with low humidity, thick shrubbery, windy weather, and lightning.**

2. How are plants in the chaparral specially designed for fire? **Seeds from many species only germinate after a fire.**

3. Should people try to put out fires that naturally occur in the chaparral? **This is a difficult question to answer. Certainly if people's property is in danger, the fires should be controlled. But studies have shown that fire is a natural part of the chaparral ecosystem and**

actually helps to keep it healthy, so many people think that natural fires should be allowed to burn when not endangering people or their property.

23. Caves

What did we learn?

1. What is a cave? **A hole or cavern inside a mountain or underground.**
2. What kinds of plants will you find in a cave ecosystem? **There are no plants inside the cave; a few may be growing near the entrance.**
3. What are the three categories of animals in a cave ecosystem? **Trogloxenes, troglophiles, and troglobites.**
4. Explain the different habits of each category of cave animal. **Trogloxenes visit the cave but do not spend their whole lives there. Troglophiles like to live in caves, but can live outside of a cave. Troglobites live their entire lives in a cave.**
5. What is the main source of nutrients in a cave ecosystem? **Bat guano.**

Taking it further

1. Why is a cave considered a low energy ecosystem? **There are no plants, so all energy must be brought in from the outside. This limits the amount of energy in the ecosystem.**
2. Why can a rise in temperature inside a cave threaten the ecosystem? **Increased temperature means increased metabolism for cold-blooded animals, requiring more food, which may not be available.**
3. Which sense is least useful in a cave? **Sight.**
4. Which senses are most useful in a cave? **Hearing, smell, and touch are more useful than sight or taste.**

24. Seasonal Behaviors

What did we learn?

1. What is hibernation? **A deep winter sleep in which the body's functions slow down greatly.**
2. What is estivation? **A deep summer sleep similar to hibernation.**
3. What is migration? **Moving from one location to another and then returning in order to survive the changing weather.**
4. List three different kinds of animals that migrate. **Birds, whales, sea turtles, butterflies and other insects, caribou, salmon, etc.**
5. What is the most likely trigger for seasonal behaviors? **The length of the day—changing number of hours of daylight.**

Taking it further

1. How can animals know where they are supposed to go when they migrate if they have never been there before? **Some animals follow their parents, but many travel by instinct.**
2. How do animals navigate while migrating? **Some use the stars, some follow scents, and others use landmarks.**
3. Why might a group of animals move from one location to another, other than for their annual migration? **Changing climate conditions or natural disasters might make food scarce, so animals will move to a new location. This is called immigration, not migration, because the animals do not usually return to the original location.**
4. If you see a monarch butterfly in the fall and then see another one in the spring, how likely is it that you are seeing the same butterfly? **It depends where you live. If you live in Mexico, you might be seeing the butterfly when it arrives and when it leaves. But if you live in Canada, it is very unlikely that the same butterfly that flew south in the fall would ever live to make it back to the north.**

25. Animal Defenses

What did we learn?

1. What are three main ways that animals try to defend themselves? **Flight, trickery, and fight.**
2. List three ways that animals can trick their enemies into leaving them alone. **Intimidation, ink, inflating their bodies, camouflage, etc.**
3. How do some eels protect themselves? **They can shock their predators with an electric pulse.**

Taking it further

1. Why do you think animals prefer to run away or frighten off enemies rather than fight? **Fighting is more dangerous. Getting away or making the enemy leave is more likely to keep the animal alive.**
2. Why do many animals prefer trickery to running away? **Trickery uses up less energy than running away.**
3. How might a defense also serve as an attack method? **An animal may use its teeth or claws to**

protect itself from its enemies and then use the same teeth and claws to attack its own prey.

26. Adaptation

What did we learn?

1. What is an adaptation? **A physical characteristic or behavior that allows an animal to survive in its environment.**

2. Are all helpful characteristics a result of a change in the organism? **No, many characteristics were part of the original created organism.**

3. What process causes different species to develop among the same kind of animal or plant? **The selection of adaptations through natural selection.**

Taking it further

1. How does natural selection work? **A kind of animal or plant can produce offspring with many different characteristics. If a particular characteristic makes an animal better suited for its environment, it will be more likely to survive and reproduce. Those offspring are more likely to have the trait that was beneficial, so they will be better suited to the environment.**

2. Does natural selection require millions of years to develop distinct populations? **No, all of the animals began reproducing after the original pairs left the Ark only a few thousand years ago and have developed into the many species we see today. There have even been observed cases of speciation.**

3. Does natural selection require genetic mutation? **No. The information for great variety was available in the original created kinds. That variety can be selected without mutations, though mutations provide more variety for selection to act on.**

How Was I Designed? Worksheet

Organism	Design features
Jack rabbit	Has large ears for greater heat dissipation in hot environments.
Woodpecker	Has toes going both directions to grasp tree; has shock absorbing skull for drilling; has sticky tongue for getting insects inside a tree.
Orchid	Has roots that can absorb water from the air so they can grow on the sides of trees where there is adequate sunlight.
Honey bee	Has pollen baskets to collect pollen when getting nectar from flowers.
Cactus	Has needles to prevent loss of water; has the ability to store large amounts of water.
Brown bat	Uses echolocation for flying and for catching insects; designed to hang upside down for long periods of time.
Oak tree	Loses its leaves in the winter.
Prairie grass	Has growing center near the ground so it can continue to grow even after being eaten over and over; goes dormant in winter.
Barn owl	Has great eyesight and hearing; has the ability to regurgitate indigestible materials.
Chameleon	Can change colors for camouflage and attracting mates.

27. Balance of Nature

What did we learn?

1. What is meant by the balance of nature? **A state in which the producers and consumers are in equilibrium.**

2. Name two ways that the balance of nature is maintained in an ecosystem. **Predator/prey feedback and territoriality are the main ways. Flocking also affects the balance.**

3. What are two ways that animals use to stake out their territory? **Singing, demonstrations, and scent markings.**

4. What happens if a male cannot find a territory to defend? **He does not mate and waits until a territory opens up.**

Taking it further

1. What would be the likely effect on the ecosystem if a prairie dog colony was devastated by the plague? **Their primary predators, the black-footed ferrets, would begin to starve and would not reproduce as quickly, lowering their population as well.**

2. What would happen if animals did not respect each others' territories? **Too many animals would breed in a given area, and there would not be enough food for everyone. Many would starve until balance was restored.**

3. How does the oxygen cycle demonstrate the balance of nature? **The amount of oxygen and carbon dioxide produced and consumed by all the plants and animals in the world is about equal, showing balance.**

4. Which methods of population control may have been present originally, and which have developed since the Fall? **Territoriality and flocking were likely**

created at the beginning; predator/prey feedback has developed since the Fall.

28. Man's Impact on the Environment

What did we learn?

1. What are some ways that farmers impact ecosystems? **They clear the land, thus changing habitats; add chemicals to their crops killing insects and weeds; use water for irrigation.**

2. What are some ways that farmers and ranchers have changed their practices to be more friendly to the environment? **Farming: contour farming, crop rotation, drip irrigation, wind breaks, GMOs, and organic farming; Ranching: introduce animals that compete with native animals for food, water, and space; sometimes kill predators.**

3. What are some ways that industry impacts ecosystems? **Using land, adding pollutants to air and water, cutting down trees, using natural resources, etc.**

Taking it further

1. What are some ways that people can minimize their impact on nature? **Recycle plastics, glass, and other items; reduce energy usage to reduce air pollution; grow plants without insecticides and herbicides.**

2. How can hunting licenses positively affect man's impact on ecosystems? **Licenses limit the number of animals that can be hunted and limit the time of year that they can be hunted. This limits man's impact on the ecosystem. Also, when certain animal populations get too large, hunting can help reduce them to sustainable sizes.**

How I Impact Nature Worksheet
Possible answers could include:

- Foods I eat: Farms take up natural areas, but crops also provide habitat. For example, birds, mice, and other animals live in corn fields or other crops.

- Transportation I use: Oil is needed to make gasoline, but oil drilling impacts ecosystems. Mining to make steel can negatively impact ecosystems, but many oil and mining operations now repair any damage done when they are finished.

- Clothes I wear: Cotton clothes are made from crops that are grown on farms; polyester clothes are made from petroleum/oil. The clothes are manufactured in factories. All of these things impact nature.

- Recreational activities: Most activities require special equipment that was manufactured at a factory that takes up space and impacts ecosystems. Fishing removes animals from the food chain. Hunting can help keep animal populations in control.

- All of these activities can be done in ways that minimize the effects on surrounding ecosystems. Also, most ecosystems quickly adapt to the "intrusion" of new farms, buildings, etc. Nature will find a new balance.

- Things that can be done to help minimize negative impacts include recycling and reusing items so fewer new items need to be manufactured, and respecting nature by not leaving trash behind when you visit a natural area. People have designed automobiles and factories that put out much less pollution than those in the past. Factories are much more careful about releasing toxic waste into the water and the air than they were in the past.

29. Endangered Species

What did we learn?

1. Name two possible natural causes of extinction of a species. **Climate change, disease, change in food supplies, etc.**

2. Name three possible man-made causes of extinction of a species. **Habitat reduction, overhunting, pollution, harassment, etc.**

3. Name three things people are doing to help endangered species. **Preserving or restoring habitat, reintroducing species, captive breeding, passing laws for protection, etc.**

Taking it further

1. Why might people overhunt a particular animal? **To make money from the animals, to stop the animals from preying on livestock, to provide food for their families, etc.**

2. Can people use the land without harming endangered species? **Since God said that man was to subdue and rule the earth, there is a way to be a good steward and still use the earth. It requires self-discipline which many people, companies, and/or nations lack.**

30. Pollution

What did we learn?

1. What is pollution? **The presence of any contaminant that harms the ecosystem.**

2. What are some natural sources of pollution? **Sand storms, volcanoes, wildfires, oil seeps, etc.**

3. What are some sources of man-made pollution? **Factories, automobiles, fireplaces, controlled burns, power plants, trash, etc.**

4. What are three major areas of the environment that can become polluted? **Air, water, and land.**

Taking it further

1. What are some ways that people can reduce water pollution? **Do not dump chemicals into the water, treat sewage, reduce the amounts of chemicals used in farming, remove oils from roads and parking lots so they are not washed away by rain.**

2. What are some ways that people can reduce air pollution? **Use cleaner burning fuels, take chemicals out of the air before it is released from a factory, do not use dangerous chemicals, don't use your fireplace as much.**

3. What are some ways that people can reduce land pollution? **Recycling, reducing the amount of wastes that are produced, reusing products.**

4. Do you think that water, air, and land are cleaner or dirtier today than they were 40 years ago? **In most parts of the world, the land, air, and water are significantly cleaner than they were 40 years ago.**

31. Acid Rain

What did we learn?

1. Why is rain naturally slightly acidic? **Water in the atmosphere combines with carbon dioxide to form carbonic acid.**

2. What is acid rain? **Rain or other precipitation that has a level of acid higher than normal rain.**

3. What are the main causes of acid rain? **Releasing of sulfur dioxide and nitrogen oxides from the burning of fossil fuels.**

4. What is buffering capacity? **The ability of the soil to neutralize acid rain.**

Taking it further

1. What are some ways to help reduce acid rain? **Reduce the sulfur dioxide and nitrogen oxides going into the air by using low sulfur coal, washing coal, scrubbing smoke, using less energy, and using alternative energy sources.**

2. If the buffering capacity were the same, would you expect acid rain to be more of a problem or less of a problem in areas with high population densities? Why? **Where there are more people, there is more need for energy, so more fossil fuels will be burned. Therefore, acid rain is more likely to be a problem in high-population areas.**

32. Global Warming

What did we learn?

1. What is the greenhouse effect? **The atmosphere traps some of the sun's energy.**

2. Why is the greenhouse effect important on earth? **It keeps the earth from being too cold.**

3. What is global warming? **The increase of the average surface temperature of the earth.**

4. What do many scientists claim are the two main causes of global warming? **Increased carbon dioxide due to the burning of fossil fuels, and deforestation.**

Taking it further

1. What are some ways that people might reduce the amount of carbon dioxide they are putting into the atmosphere? **Use less energy, change to alternative energy sources that do not use fossil fuels, and reduce deforestation.**

2. Why is it inappropriate to panic about global warming? **First, we are all in God's hands, and we must trust Him and not panic. Second, the data is not conclusive, and the models are not reliable.**

33. What Can You Do?

What did we learn?

1. What are the three R's of conservation? **Reduce, reuse, and recycle.**

2. List two ways you plan to do each of these things. **Answers will vary.**

Taking it further

1. Why is it important to be concerned about how humans impact the environment? **God has given people the job of caring for His creation, and we need to take that job seriously.**

Properties of Atoms & Molecules → Worksheet Answer Keys

1. Introduction to Chemistry

What did we learn?

1. What is matter? **Anything that has mass and takes up space.**

2. Does air have mass? **Yes. It may seem like there is nothing there, but even though air is very light, it still has mass. The air contains molecules that take up space.**

3. What do chemists study? **The way matter reacts with other matter and the environment.**

Taking it further

1. Would you expect to see the same reaction each time you combine baking soda and vinegar? **Yes, because God designed certain laws for matter to follow, we would expect it to react the same way each time.**

2. Atoms

What did we learn?

1. What is an atom? **The smallest part of matter that cannot be broken down by ordinary chemical means.**

2. What are the three parts of an atom? **Protons, neutrons, and electrons.**

3. What electrical charge does each part of the atom have? **Protons are positive, neutrons are neutral, and electrons are negative.**

4. What is the nucleus of an atom? **The dense center of the atom consisting of protons and neutrons.**

5. What part of the atom determines what type of atom it is? **The number of protons in the nucleus determines what kind of atom it is.**

6. What is a valence electron? **An electron in the outermost energy level for that atom.**

Taking it further

1. Why is it necessary to use a model to show what an atom is like? **Atoms are too small to see and are very complex, so a model is useful for understanding what an atom is like.**

2. On your worksheet, you colored neutrons blue and protons red. Are neutrons actually blue and protons actually red in a real atom? **No, the colors used in a model are just to help us visualize the parts. They do not really represent the actual colors.**

Atomic Models Worksheet

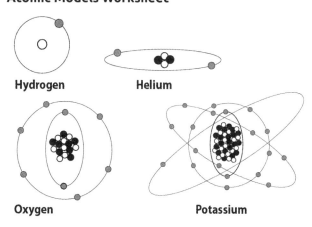

Hydrogen   Helium

Oxygen   Potassium

Challenge: Energy Levels Worksheet

See page 433.

3. Atomic Mass

What did we learn?

1. What are the three particles that make up an atom? **Proton, electron, and neutron.**

2. What is the atomic number of an atom? **The number of protons in the nucleus.**

3. What is the atomic mass of an atom? **The sum of the protons and neutrons in the nucleus of the atom.**

4. How can you determine the number of electrons, protons, and neutrons in an atom if you are given the atomic number and atomic mass? **The number of protons is the same as the atomic number. The number of electrons is equal to the number of protons. The number of neutrons is equal to the atomic mass minus the number of protons.**

Taking it further

1. What does a hydrogen atom become if it loses its electron? **A proton.**

2. Why are electrons ignored when calculating an element's mass? **The mass of an electron is so small compared to the mass of a proton or neutron that it does not make a significant difference.**

Learning About Atoms Worksheet
See page 433.

Challenge: Understanding Atoms Worksheet
See page 433.

4. Molecules

What did we learn?
1. What is a molecule? **Two or more atoms chemically connected or bonded together.**
2. What is a diatomic molecule? **A molecule with two of the same type of atoms connected together.**
3. What is a compound? **A molecule made from two or more different kinds of atoms.**

Taking it further
1. What is the most important factor in determining if two atoms will bond with each other? **The number of valence electrons each atom has.**
2. Table salt is a compound formed from sodium and chlorine. Would you expect sodium atoms and chlorine atoms to taste salty? Why or why not? **No, because when molecules are formed, the resulting compound is a new substance with its own characteristics, completely different from those of the original elements.**

What Am I? Worksheet
Gold (Au): **Element**
Ammonia (NH_3): **Compound**
Oxygen (O_2): **Diatomic molecule**
Nitrogen (N_2): **Diatomic molecule**
Silver (Ag): **Element**
Salt (NaCl): **Compound**
Sucrose ($C_{12}H_{22}O_{11}$): **Compound**
Helium (He): **Element**
Water (H_2O): **Compound**
Baking soda ($NaHCO_3$): **Compound**

5. Periodic Table of the Elements

What did we learn?
1. How many valence electrons do the elements in each column have? **Those in column IA have 1 valence electron; those in column IIA have 2, IIIA have 3, etc. Those in columns IB–VIIIB have varying numbers of valence electrons. To determine this is beyond the scope of this course.**
2. What four pieces of information are included for each element in any periodic table of the elements? **The element name, symbol, atomic number, and atomic mass.**
3. What do all elements in a column on the periodic table have in common? **They have the same number of valence electrons.**
4. What do all elements in a row on the periodic table have in common? **They have electrons in the same number of energy levels/same number of electron layers.**

Taking it further
1. Atoms are stable when they have eight electrons in their outermost energy level. Therefore, elements from column IA will react easily with elements from which column? **Column VIIA.**
2. Elements from column IIA will react easily with elements from which column? **Column VIA.**

Learning About the Elements Worksheet
1. What is the symbol for calcium? **Ca.**
2. What is the symbol for silver? **Ag.**
3. What is the atomic number for copper? **29.**
4. What is the atomic mass for rutherfordium? **261 (atomic number 104).**
5. What are two elements in the same column as sodium? **H, Li, K, Rb, Cs, Fr.**
6. What are two elements with eight electrons in their outer layer? **Ne, Ar, Kr, Xe, Rn. (All elements in column VIIIA except Helium, which has only 2 valence electrons.)**
7. How many electrons are in the outer layer of nitrogen? **5. (It is in column VA.)**
8. How many layers of electrons does barium have? **6. (It is in row 6.)**
9. Name one transition element. **Accept anything from columns IB–VIIB.**
10. Would silicon be more likely to react the same way as carbon or chlorine? **Carbon—they are in the same column and have the same number of valence electrons.**

6. Metals

What did we learn?
1. What are the six characteristics of most metals? **Silvery luster, solid, malleable, ductile, conducts electricity, reacts with other elements.**

2. How many valence electrons do most metals have? **Most commonly, metals have 1 or 2 valence electrons, but some have 3 or 4.**

3. What is a metalloid? **An element that has some metal characteristics and some nonmetal characteristics.**

Taking it further

1. What are the most likely elements to be used in making computer chips? **The semiconductors—the ones shaded dark green on the periodic table of the elements. The most commonly used elements are silicon, germanium, and boron.**

2. Is arsenic likely to be used as electrical wire in a house? **No, it is only a semiconductor, so it would not make good electrical wiring.**

7. Nonmetals

What did we learn?

1. What are some common characteristics of nonmetals? **Not shiny or silver, not conductive, do not easily lose electrons.**

2. What is the most common state, solid, liquid, or gas, for nonmetal elements? **Gas.**

3. Why are halogens very reactive? **They need only one electron to fill their outer shells.**

4. Why are noble gases very non-reactive? **They have a full outer shell of electrons.**

Taking it further

1. Hydrogen often acts like a halogen. How might it act differently from a halogen? **Because hydrogen has only one electron, it can give up its electron and become an ion, whereas halogens do not easily give up electrons.**

2. Why are balloons filled with helium instead of hydrogen? **Helium is a noble gas and non-reactive, but hydrogen is highly reactive. We don't want balloons exploding as the hydrogen reacts with another element.**

8. Hydrogen

What did we learn?

1. What is the most common element in the universe? **Hydrogen.**

2. What is the atomic structure of hydrogen? **It has one proton and one electron.**

3. What is the atomic number for hydrogen? **1.**

4. Why is hydrogen sometimes grouped with the alkali metals? **It has only one electron, so it often behaves like an alkali metal.**

5. Why is hydrogen sometimes grouped with the halogens? **It is stable if it gains one electron, so it often behaves like a halogen.**

Taking it further

1. Why is hydrogen one of the most reactive elements? **Most elements must either gain electrons or lose electrons to combine with other elements. Hydrogen can do either one, so it combines easily with many other elements.**

2. Margarine contains only partially hydrogenated oil. What do you suppose fully hydrogenated oils are like? **They are much harder or more solid than margarine and are not easily spread.**

9. Carbon

What did we learn?

1. What is the atomic number and atomic structure of carbon? **Carbon is element number 6. It has 6 protons, 6 neutrons, and 6 electrons.**

2. What makes a compound an organic compound? **It contains carbon atoms.**

3. Name two common forms of carbon. **Graphite and diamond.**

4. What is one by-product of burning coal? **Carbon dioxide. (Water is also a by-product but was not mentioned in the lesson.)**

Taking it further

1. How does the carbon cycle demonstrate God's care for His creation? **It allows carbon to be recycled and keeps life continuing on earth.**

2. What is the most likely event that caused coal formation? **The Genesis Flood would have buried large amounts of plants under tons of mud and water. This is the most likely cause of the large amounts of coal found in the earth.**

3. What would happen if bacteria and fungi did not convert carbon into carbon dioxide gas? **The carbon from dead plants and animals would become trapped and would not be able to be reused in the growth of new plants.**

10. Oxygen

What did we learn?

1. What is the atomic structure of oxygen? **Oxygen has 8 protons and 8 neutrons in the nucleus and 8 electrons. It has 6 valence electrons.**
2. How is ozone different from the oxygen we breathe? **Ozone is a molecule of three oxygen atoms. The oxygen we breathe is a molecule of two oxygen atoms. O_3 is poisonous and O_2 is not.**

Taking it further

1. Why does the existence of ozone in the upper atmosphere show God's provision for life on earth? **If ozone were in the lower atmosphere, it would poison all living things. But in the upper atmosphere, it protects the earth from harmful radiation.**
2. How do animals in the ocean get the needed oxygen to "burn" the food they eat? **Most aquatic animals have gills that extract oxygen from the water. A few, like whales and dolphins, have to surface and breathe air.**
3. Why are oxygen atoms nearly always combined with other atoms? **They have only six valence electrons, so they are not stable by themselves.**

11. Ionic Bonding

What did we learn?

1. What is the main feature in an atom that determines how it will bond with other atoms? **The number of valence electrons it contains.**
2. What kind of bond is formed when one atom gives up electrons and the other atom takes the electrons from it? **An ionic bond.**
3. What is electronegativity? **A measure of how strongly an element attracts electrons to itself.**
4. Why are compounds that are formed when one element takes electrons from another called ionic compounds? **Because ions are formed when electrons are taken away or added.**
5. What are some common characteristics of ionic compounds? **Conduct electricity when melted or dissolved, high melting point, soluble in water, brittle, form ions, form crystal lattices.**
6. Which element has a higher electronegativity, chlorine or potassium? **Electronegativity increases as you go from left to right across the periodic table. Chlorine holds on to its electrons more tightly than potassium, so it has a higher electronegativity.**

Taking it further

1. Which column of elements are the atoms in column IA most likely to form ionic bonds with? **The elements in column VIIA.**
2. Use the periodic table of the elements to determine the number of electrons that barium would give up in an ionic bond. **Barium is in column IIA, so it has 2 valence electrons that it would give up.**

Challenge: Name That Ion Worksheet

NaF – **Sodium fluoride**   LiBr – **Lithium bromide**
KCl – **Potassium chloride**   CaS – **Calcium sulfide**
$CaCl_2$ – **Calcium chloride**

12. Covalent Bonding

What did we learn?

1. What is a covalent bond? **A bond formed when electrons are shared between two or more atoms.**
2. What are some common characteristics of covalent compounds? **Do not conduct electricity, low melting point, strong, flexible, lightweight, insoluble in water, only slight attraction for each other.**
3. What is the most common covalent compound on earth? **Water.**

Taking it further

1. Why do diatomic molecules form covalent bonds instead of ionic bonds? **Diatomic molecules are formed from two atoms of the same element, so they have the same electronegativity. Since neither atom is able to take away or give up its electrons, they cannot form ionic bonds.**
2. Would you expect more compounds to form ionic bonds or covalent bonds? **Since there are so many metals and only a few metalloids and nonmetals, you might expect most compounds to be ionic. However, there are actually so many different ways to share electrons that covalent bonds are actually more common.**

Challenge: Bonding Experiment Worksheet

Water–H_2O is composed of _**nonmetals**_
Baking soda–$NaHCO_3$ is composed of _**both**_
Sugar (sucrose–$C_{12}H_{22}O_{11}$) is composed of _**nonmetals**_
Salt–NaCl is composed of _**both**_
Olive oil (oleic acid–$C_{18}H_{34}O_2$) is composed of _**nonmetals**_

Substances that are formed from all nonmetals are covalent. Substances formed from at least one metal and one nonmetal are ionic.

Compound tested	Did it conduct electricity? (Observations)	Ionic or covalent? (Conclusions)
Distilled water	No	Covalent
Baking soda	Yes	Ionic
Sugar	No	Covalent
Salt	Yes	Ionic
Olive oil	No	Covalent

13. Metallic Bonding

What did we learn?

1. What is the free electron model? **It is the theory that metals form bonds by sharing electrons on a very large scale. Thousands of atoms allow their electrons to freely move about so that the atoms remain stable.**

2. How many valence electrons do metals usually have? **Usually 1, 2, or 3.**

3. What are common characteristics of metallic compounds? **Free electrons, conduct electricity and heat, shiny luster, high melting point, insoluble in water.**

Taking it further

1. Why don't metals form ionic or covalent bonds? **Because they have similar numbers of valence electrons, they do not pull electrons away from each other. Also, because they have a low number of valence electrons, they do not have enough to share among a small number of atoms. Therefore, they must share on a large scale—among thousands of atoms.**

2. Would you expect semiconductors to form metallic bonds? **No. Since they do not conduct electricity well, they would not have free electrons.**

Challenge: Bonding Characteristics Worksheet

Ionic bonding	Covalent bonding	Metallic bonding
A	C	B
F	E	D
G	H	I
J	K	J
L	M	L
O	N	N

14. Mining & Metal Alloys

What did we learn?

1. What element is combined with most metals to form metal ore? **Most metals are in the form of metal oxides—metals combined with oxygen.**

2. What must be done to metal oxides to obtain pure metal? **The oxygen must be removed through a reduction reaction.**

3. What is an alloy? **A metal that has a small amount of another metal added to it.**

4. Why are alloys produced? **Alloys are often stronger, more resilient, and easier to work with than pure metals.**

Taking it further

1. Do you think chromium would be added to steel that is going to be used in saw blades? Why or why not? **Probably not. Chromium keeps steel from oxidizing; however, a little oxidation on a saw blade will not keep it from working. The saw blade needs to be strong, so tungsten may be added, but not chromium.**

2. Is oxidation of metal always a bad thing? **Not always. Sometimes a layer of oxidation prevents more oxygen from reaching the rest of the metal. So, leaving a small amount of oxidation can actually reduce the overall amount of oxidation that occurs. This is why the green layer is not cleaned off of the Statue of Liberty. This is not always the case, however; sometimes oxidation, such as rust, continues to occur until the sample is completely gone.**

15. Crystals

What did we learn?

1. What is a crystal? **A substance whose atoms are lined up in a regular lattice configuration. Crystals have smooth faces and defined edges.**

2. How do crystals form? **When a liquid cools slowly, the atoms line up in regular patterns to form crystal lattices based on their chemical characteristics.**

3. What is an artificial gem? **One that is formed by man and not formed naturally.**

4. Where would you look to find crystals? **In rocks and minerals, in the kitchen (salt and sugar), in caves, jewelry.**

Taking it further

1. Why are naturally occurring gems more valuable than artificial gems when many are made from the same materials? **Even though they are made**

from the same materials, artificial gems do not have the same strength and brilliance of naturally occurring crystals. God's crystals are still better than man's.

2. Why is a saturated solution better for forming crystals? **The more atoms of the crystal-forming material you have, such as salt, the more likely they are to line up in a lattice formation.**

3. What are some ways you use crystals in your home? **In food, in your computer, TV, phone, and other electronic devices, in your rock collection, gems in your mother's wedding ring, etc.**

16. Ceramics

What did we learn?

1. What is ceramic? **It is a material that is formed when ingredients fuse together by heat, often made with clay.**

2. What are some examples of traditional ceramics? **Pottery, brick, porcelain, and glass.**

3. What makes ceramics hard? **The material forms crystals when it is baked or fired.**

4. What are some advantages of modern ceramics? **They are hard, strong, heat resistant, and don't rust.**

Taking it further

1. Why are the tiles on the space shuttle made of ceramic? **Because ceramic is very heat resistant, the tiles keep the heat generated by friction with the atmosphere away from the shuttle, allowing the shuttle to reenter the atmosphere without burning up.**

2. Why are crystalline structures stronger than noncrystalline structures? **The lattice shape of the bonds allows atoms to be connected in more than one direction, so the compounds are stronger.**

17. Chemical Reactions

What did we learn?

1. What is a chemical reaction? **When atomic bonds are formed or broken, when two or more elements combine together to form a new substance, or when a substance is broken down into its separate elements.**

2. What are the initial ingredients in a chemical reaction called? **Reactants.**

3. What are the resulting substances of a chemical reaction called? **Products.**

Taking it further

1. How might you speed up a chemical reaction? **Add heat; add surface area to the reactants by changing their shape—make them thinner or break or crush them; increase the concentration of the reactants; add a catalyst.**

2. A fire hose usually sprays water on a fire to put it out. Water does not deprive the fire of oxygen, so why does water put out a fire? **Water absorbs the heat from the fire, and heat is another necessary ingredient in producing and sustaining a fire.**

3. What chemical reaction do you think is taking place in the making of a loaf of bread? **The yeast reacts with the sugar in the bread dough to produce carbon dioxide.**

Challenge: Reaction Rate Experiment Worksheet

You should see the tablet in the hot water dissolve more quickly than the tablets in the other cups. You should see the crushed tablet dissolve more quickly than the tablets in the other cups. Usually, the higher the temperature, the faster the reaction will take place. In general, the greater the surface area of the reactants, the faster the reaction will take place.

18. Chemical Equations

What did we learn?

1. What is a chemical equation? **It is an equation that visually shows what happens to each element in a chemical reaction.**

2. What are the elements or compounds on the left side of a chemical equation called? **The reactants.**

3. What are the elements or compounds on the right side of a chemical equation called? **The products.**

Taking it further

1. Why is it helpful to use chemical equations? **Equations provide a visual way to see what is happening in a chemical reaction without drawing pictures.**

Understanding Chemical Equations Worksheet

1. $C + O_2 \longrightarrow CO_2$
2. $N_2 + 3H_2 \longrightarrow 2NH_3$
3. $2H_2O \longrightarrow 2H_2 + O_2$

Challenge: Reactants & Products Worksheet

1. $4Al + 3O_2 \longrightarrow$ _B_

2. $H_2SO_4 + 2\ LiOH \longrightarrow$ _A_
3. $4\ NH_3 + 3\ O_2 \longrightarrow$ _C_
4. $P_4 + 10\ Cl_2 \longrightarrow$ _E_
5. $CO_2 \longrightarrow$ _D_
6. $H + OH \longrightarrow$ _F_
7. $2\ KClO_3 \longrightarrow$ _H_
8. $2\ Na + 2\ H_2O \longrightarrow$ _G_

- Which of the above equations represent decomposition reactions? _ **5, 7**_
- Which of the above equations represent composition reactions? _**1, 4, 6**_
- Which of the above equations represent single displacement reactions? _ **3, 8**_
- Which of the above equations represent double displacement reactions? _**2**_

19. Catalysts

What did we learn?

1. What is a catalyst? **A substance added to speed up a chemical reaction.**
2. How does a catalyst work? **It reduces the amount of energy needed for the chemical reaction to take place.**
3. What is an inhibitor? **A substance that slows down or prevents a chemical reaction.**
4. What is an enzyme? **A catalyst found in living cells.**

Taking it further

1. Why is it important that living cells have enzymes? **If enzymes were not available, many chemical reactions such as digestion would take much too long to occur.**
2. Are catalysts always good? **Not necessarily. If a catalyst caused food to spoil very quickly, that would be a bad use of a catalyst.**

20. Endothermic & Exothermic Reactions

What did we learn?

1. What is an exothermic reaction? **A chemical reaction that releases energy.**
2. What is an endothermic reaction? **A chemical reaction that absorbs energy.**

Taking it further

1. If a chemical reaction produces a spark, is it likely to be an endothermic or exothermic reaction? **Light is a form of energy, so it would be an exothermic reaction.**
2. How do photosynthesis and digestion reveal God's plan for life? **Photosynthesis absorbs and stores energy from the sun in the sugar molecules in the plant. That energy is released during digestion after an animal eats the plant. This is God's plan for providing necessary food, and therefore energy, for all of the animals—and humans—on earth.**
3. If the temperature of the product is lower than the temperature of the reactants, was the reaction endothermic or exothermic? **If the result is cooler than the beginning reactants, then energy was absorbed, so the reaction was endothermic.**

Challenge: Endothermic or Exothermic? Worksheet

The results should show the reaction is endothermic; the temperature of the water goes down during the reaction then levels off.

21. Chemical Analysis

What did we learn?

1. What is chemical analysis? **Using chemical reactions to determine the composition of a substance.**
2. List three different types of chemical analysis. **Flame test, spectrometer, indicators.**
3. What is a chemical indicator? **A substance that changes color when it reacts with a specific chemical.**
4. What is the pH scale? **The scale used to measure the strength of an acid or a base.**
5. What does a pH of 7 tell you about a substance? **It is neutral. It is not an acid or a base.**

Taking it further

1. Why is it important to periodically test the pH of swimming pool water? **Water must be close to neutral to be safe to swim in. Also, water with a pH much greater than 6.8–7.0 can cause pipes to become clogged with minerals.**
2. Name at least one other use for testing pH of a liquid. **Hair treatments like permanents must be tested for pH so that hair curls and doesn't burn. Urine can be tested for pH to detect health problems. Beverages are tested for proper pH to ensure proper taste. Drinking**

water is tested for proper pH, and wastewater is tested before releasing it back into the water system.

22. Acids

What did we learn?

1. What defines a substance as an acid? **It produces hydronium ions when dissolved in water.**
2. What is a hydronium ion? **H_3O^+, formed by a water molecule and a hydrogen ion.**
3. How is a weak acid different from a strong acid? **A weak acid holds onto its hydrogen atoms more strongly than a strong acid, so it forms fewer hydronium ions in water.**
4. What are some common characteristics of an acid? **Sour taste, conducts electricity in water, reacts with metals, many are corrosive, neutralizes bases, reacts with indicators.**
5. How can you tell if a substance is an acid? **Dissolve it in water and use an indicator to test for acid.**

Taking it further

1. Why is saliva slightly acidic? **The acid in your saliva helps begin the digestion process by helping break down the food molecules.**
2. Would you expect water taken from a puddle on the forest floor to be acidic, neutral, or basic? Why? **It would probably be acidic because the forest floor is covered with decaying plants, and decaying plants produce humic acid.**
3. What would you expect to be a key ingredient in sour candy? **Some kind of acid. Sour spray and other sour candies often contain several types of acids.**

23. Bases

What did we learn?

1. What defines a substance as a base? **It produces hydroxide ions when dissolved in water.**
2. What is a hydroxide ion? **OH^- ion. Essentially, it is a molecule containing one oxygen atom and one hydrogen atom, but it contains an extra electron, giving it a negative charge, so it is considered an ion instead of a molecule.**
3. How is a weak base different from a strong base? **A weak base holds onto its hydroxide ions more strongly than a strong base does.**
4. What are some common characteristics of a base? **Bitter taste, conducts electricity in water, feels slippery, many are corrosive, neutralizes acids, reacts with indicators.**

5. How can you tell if a substance is a base? **Dissolve it in water and use an indicator to test for base.**

Taking it further

1. If you spill a base, what should you do before trying to clean it up? **Add an acid to neutralize it.**
2. Do you think that strontium (Sr) is likely to form a strong base? Why or why not? **Strontium is in the alkali metal family, and alkali metals tend to form strong bases. Therefore, strontium is likely to form a strong base.**

24. Salts

What did we learn?

1. How is a salt formed? **When a negative acid ion combines with a positive base ion, a salt is formed.**
2. What are two common characteristics of salts? **They have a salty flavor, and they form crystals.**
3. How are salt families named? **By the acid from which they are made.**
4. Name three salt families. **Sulfates, chlorides, nitrates, carbonates, phosphates, potash.**

Taking it further

1. What do you expect to be the results of combining vinegar and lye? **You would get salt and water.**
2. Why are some salts still acidic or basic? **The ions do not completely combine together, so some hydrogen or hydroxide ions are still present.**

Challenge: Acid/Base Reactions Worksheet

1. $HClO_3 + KOH \longrightarrow KClO_3 + H_2O$

The acid is **$HClO_3$** The salt is **$KClO_3$**

The base is **KOH**

2. $HBr + Ca(OH)_2 \longrightarrow CaBr_2 + H_2O$

The acid is **HBr** The salt is **$CaBr_2$**

The base is **$Ca(OH)_2$**

3. $H_2SO_4 + 2NH_3 \longrightarrow 2NH_4^+ + SO_4^{2-}$

The acid is **H_2SO_4**

The base is **NH_3**

4. $HI + H_2O \longrightarrow H_3O^+ + I^-$

The acid is **HI**

The base is **H_2O**

25. Biochemistry

What did we learn?

1. List at least two chemical functions performed inside living creatures. **Plants perform photosynthesis—turning carbon dioxide and water into sugar and oxygen. Plants and animals perform cellular respiration—turning sugar and oxygen into carbon dioxide and water. In many animals, oxygen combines with hemoglobin for easy transport throughout the body.**

2. What is the chemical reaction that takes place during photosynthesis? **Water and carbon dioxide chemically combine to form sugar and oxygen.**

3. What is the main chemical reaction that takes place during digestion? **Sugar and oxygen chemically combine to form carbon dioxide and water; also, larger molecules are broken down into smaller molecules.**

4. What substance is necessary for nearly every chemical reaction in living things? **Water.**

5. Name the three major chemicals your body needs that are found in the foods we eat. **Proteins, fats, and carbohydrates.**

Taking it further

1. Why did God design your body to have enzymes? **Enzymes help digestion and other metabolic processes to occur at a much quicker rate than they otherwise would.**

2. With what you know about chemical processes, why do you think it is important to brush your teeth after you eat? **The chemicals in your mouth begin the digestion process. These chemicals can cause tooth decay if they stay in your mouth too long. So you need to brush away any food and acids so your teeth stay healthy.**

3. Can you think of other chemical processes in your body besides the ones mentioned in this lesson? **Taste and smell are chemical reactions; many chemical reactions take place to cause blood to clot; chemical reactions take place to make your nerves send messages. The list is nearly endless.**

Challenge: Enzyme Reaction Worksheet

1. What effects did the plain pineapple juice have on the gelatin? **The plain pineapple juice should break down the protein in the gelatin, preventing the gelatin from gelling.**

2. What effects did changing the pH have on the way the juice affected the gelatin? **The vinegar changes the pH of the juice to be acidic, which damages the protease, preventing it from breaking down the protein, so the gelatin should still be able to gel.**

3. What effects did heating the juice have on the way the juice affected the gelatin? **The heat damages the protease so the gelatin should still be able to gel.**

4. Why was cup 4 necessary? **Cup 4 is the control. It shows what happens when nothing special is done. All of the other cups can be compared to this cup to show what effect each change had on the gelatin.**

26. Decomposers

What did we learn?

1. What is a scavenger? **An animal that eats dead animals.**

2. What is a decomposer? **An organism that breaks down dead plants, dead animals, or dung into simple chemical compounds.**

3. What is this way of recycling nitrogen called? **The nitrogen cycle.**

Taking it further

1. Why are decomposers necessary? **They are needed to break down complex compounds into simple compounds that can be used by plants. Without decomposers, the elements would be locked up and plants would not be able to grow.**

2. Were there animal scavengers in God's perfect creation, before the Fall of man? **No, there was no animal or human death before Adam sinned. Man and animals were all created to be vegetarians—see Genesis 1:27–31.**

3. Explain how a compost pile allows you to participate in the nitrogen cycle. **You can take food scraps, such as potato peels, and place them in a bin or pile outside. Bacteria or other decomposers eat these scraps, leaving behind compost, which is nutrient-rich material that you can add to your garden. You have taken nitrogen from the food scraps and returned it to the soil to be used by the plants you grow in your garden.**

27. Chemicals in Farming

What did we learn?

1. What are three ways that farmers ensure their soil will have enough nutrients for their crops? **Adding fertilizers, allowing the land to lie fallow, crop rotation, burning of unwanted plants.**

2. What is hydroponics? **Growing plants without soil, using chemicals in water.**

3. How are chemicals used in farming other than for nutrients for the plants? **Chemicals are used to kill pests, diseases, and unwanted plants—pesticides, fungicides, and herbicides.**

4. How is an organic farm different from other farms? **Organic farms do not use man-made chemicals.**

Taking it further

1. Why did the farmers let cattle graze on their land once every fourth year in the Norfolk 4-course plant rotation method? **The animal waste added nutrients back into the soil.**

2. How does hydroponics replace the role of soil in plant growth? **A framework is provided to support the plants, and nutrients are added to the water for absorption by the roots.**

28. Medicines

What did we learn?

1. Why are chemicals used as medicines? **Your body is constantly performing chemical reactions, so adding chemicals to your body causes different reactions to occur.**

2. What were the earliest recorded medicines? **Herbs.**

3. What was Sir Alexander Fleming's important discovery? **Penicillin—the first antibiotic.**

Taking it further

1. If plants have the potential of supplying new medicines, where might a person look to find different plants? **One of the likeliest sources of medicinal plants is believed to be the tropical rainforests where there are thousands of unusual plants.**

2. What other sources might there be for discovering new medicines? **In addition to plants, animals in the rainforest and ocean are likely places to test for new medicines. Also, a better understanding of how the human body processes chemicals can lead to the development of new synthetic medicines.**

29. Perfumes

What did we learn?

1. What is a perfume? **A liquid with a pleasing smell.**

2. What must be removed from flower petals to make perfume? **The fragrant oil.**

3. Describe the two main methods for removing oil from flower petals. **With solvent extraction, a solvent is used to dissolve the oils, then the solvent is allowed to evaporate. With steam distillation, steam is used to vaporize the oil, then both the oil and water condense, and the oil is skimmed off the top of the water.**

Taking it further

1. Why should you test a new perfume on your skin before you buy it? **The scent of the perfume in the bottle may not be the same as it is on your skin. The alcohol in the bottle may mask the true scent. So put some on your skin and see how it smells once the alcohol has evaporated.**

2. Why wasn't it necessary to use one of the methods described in the lesson to make your homemade perfume? **As the cloves soaked in the alcohol, the scent particles slowly moved into the alcohol from the cloves. This is a very slow process. The methods described in the lesson greatly speed up the process for commercial production of perfume.**

30. Rubber

What did we learn?

1. What is natural rubber made from? **Latex from a rubber tree.**

2. What is synthetic rubber made from? **Petroleum—oil.**

3. What is vulcanization? **The process of adding sulfur to rubber to make it elastic in all types of weather.**

4. What is a polymer? **A long chain of molecules connected together.**

Taking it further

1. Why is it difficult to recycle automobile tires? **The vulcanization process makes the rubber very long-lasting, but it also makes it hard to break down the molecules, so recycling is difficult.**

2. What advantages and disadvantages are there to using synthetic rubber instead of natural rubber? **Synthetic rubber is cheaper than natural rubber; however, it requires petroleum, much of which America must import from other countries.**

31. Plastics

What did we learn?

1. What is plastic? **A substance made from polymers that are derived from petroleum.**

2. What was celluloid, the first artificial polymer, made from? **From cellulose that comes from cotton plants.**

3. What is the difference between thermoplastic and thermosetting resin? **Thermoplastics will become soft when reheated, thermosetting resin plastic will not.**

4. Why are people concerned about throwing plastic items away? **Plastic does not decompose easily.**

5. What does the recycling number on a plastic item mean? **It represents the type of plastic the item is made of.**

6. Why are plastic bags usually recycled separately from other plastics? **They get caught in the machinery that separates recyclables.**

Taking it further

1. Name three ways that plastic is used in sports. **Plastic or vinyl balls, artificial rubber soles on running shoes, plastic hooks to hold soccer nets in place, polyester sports clothes, and many other uses.**

2. What advantages do plastic items have over natural materials? **Many plastic items are stronger, more flexible, and longer lasting than their natural counterparts.**

Chemical Word Search

32. Fireworks

What did we learn?

1. What are the key ingredients in a fireworks shell? **The chemical that releases the light, black powder for the explosion, and fuses to light the powder.**

2. Why does a fireworks shell have two different black powder charges? **One charge lifts the shell into the air, and the other charge blasts the shell open.**

3. How do fireworks generate flashes of light? **When the blasting charge explodes, the energy released forces electrons in the chemicals into higher energy levels. When the electrons return to their normal energy levels, they release energy in the form of light.**

4. What determines the color of the firework? **The chemical compound that is packed inside.**

Taking it further

1. How can a firework explode with one color and then change to a different color? **Two different chemicals are packed in the shell and ignited at different times.**

2. Why would employees at a fireworks plant have to wear only cotton clothing? **Nylon, polyester, silk, and other fabrics can build up a static charge. This could be very dangerous when working around black powder because a static discharge could ignite the powder.**

33. Rocket Fuel

What did we learn?

1. What is combustion? **A chemical reaction that produces great amounts of heat.**

2. What two elements are combined in most modern rocket fuel? **Oxygen and hydrogen.**

3. What compound is produced in this reaction? **Water/steam.**

4. How does combining oxygen and hydrogen produce lift? **The reaction takes place at very high temperatures, heating the atoms to very high temperatures and thus very high speeds. These molecules exit the engine at great speeds, thus producing lift because of Newton's third law of motion.**

5. What is Newton's third law of motion? **For every action there is an equal and opposite reaction.**

Taking it further

1. Why is oxygen and hydrogen a better choice for rocket fuel than kerosene was? **The end product of the reaction of oxygen and hydrogen is steam, and the end product of kerosene combustion is carbon dioxide. Water is lighter than carbon dioxide, so it can move faster. The faster the molecules are moving when they leave the rocket engine, the more lift they produce.**

34. Fun with Chemistry: Final Project

What did we learn?
1. What was your favorite chemical reaction? **Answers will vary.**
2. Why did you like that reaction? **Answers will vary.**

Taking it further
1. What do you think will happen if you use skim milk in the first activity? **There are very few fat molecules in skim milk, so adding the soap will make little difference. The colors will eventually mix, but at a much slower rate.**
2. What colors would you expect to see separate out of orange ink? Brown ink? **Orange is a combination of yellow and red. Brown is a combination of yellow, red, and blue.**
3. Why is it important not to inhale the sodium polyacrylate from the diaper? **Evan a small amount of this chemical will absorb a lot of water, so it can irritate your lungs and your eyes by drying them out.**

Challenge: River Facts Sheet

River	Length	Size of river basin	Discharge at mouth	Countries or states it flows through	Major tributaries
Amazon	3920 mi 6308 km	2,270,000 sq. mi 7,050,000 sq. km	219,000 cu. meters/sec.	Peru, Brazil, Columbia, Venezuela, Bolivia	Negro, Tocantins
Congo	2900 mi 4700 km	1,440,000 sq. mi 3,822,000 sq. km	42,000 cu. meters/sec.	African Republic, Republic of the Congo, Angola, Zambia, Tanzania	Ubangi River, Aruwimi, Kasai, Lomami
Nile	4180 mi 6727 km	1,312,000 sq. mi 3,40,000 sq. km	2830 cu. meters/sec.	Ethiopia, Sudan, Egypt, Rwanda, Tanzania, Uganda, Burundi, Dem, Rep. of Congo, Eritrea, Kenya	White Nile, Blue Nile
Mississippi	2320 mi 3734 km	1,151,000 sq. mi 2,981,000 sq. km	12,743 cu. meters/sec.	USA: MN, WI, IA, IL, MO, KY, TN, AR, LA, MS	Ohio, Missouri, Arkansas, Tennessee
Yangtze	3964 mi 6379 km	680,000 sq. mi 1,970,000 sq. km	35,000 cu. meters/sec.	China	Yalong, Minjiang, Jialing, Tuo he, Han
Rio De La Plata	2795 mi 4500 km	1,197,000 sq. mi 3,100,000 sq. km	17,100 cu. meters/sec.	Argentina, Uruguay	Paraguay, Pilcomayo, Parana, Uruguay
Hwang Ho/ Yellow	3395 mi 5464 km	290,000 sq. mi 745,000 sq. km	2,571 cu. meters/sec.	China	White, Black, Huang River
Orinoco	1300 mi 2100 km	340,000 sq. mi 880,000 sq. km	33,000 cu. meters/sec.	Venezuela, Brazil	Apure, Caura, Caroni
Yukon	2200 mi 3685 km	330,000 sq. mi 855,000 sq. km	6,430 cu. meters/sec.	United States (Alaska), Canada	Pelly, Porcupine Tanana
Volga	2290 mi 3688 km	533,000 sq. mi 1,380,000 sq. km	8,000 cu. meters/sec.	Russia	Kama, Oka, Moskva

Challenge: Energy Levels Worksheet

Element	Energy levels	Electrons in level 1	Electrons in level 2	Electrons in level 3	Electrons in level 4	Electrons in level 5	Electrons in level 6
He Helium	1	2					
Be Beryllium	2	2	2				
Al Aluminum	3	2	8	3			
Cl Chlorine	3	2	8	7			
Fe Iron	4	2	8	14	2		
Kr Krypton	4	2	8	18	8		
Ag Silver	5	2	8	18	18	1	
Au Gold	6	2	8	18	32	18	1

Learning About Atoms Worksheet

Element	Atomic number	Atomic mass	# of protons	# of electrons	# of neutrons
Hydrogen	1	1	1	1	0
Helium	2	4	2	2	2
Oxygen	8	16	8	8	8
Fluorine	9	19	9	9	10
Chromium	24	52	24	24	28

Challenge: Understanding Atoms Worksheet

Element	Symbol	Atomic number	Atomic mass	# of protons	# of electrons	Most common # of neutrons
Hydrogen	H	1	1.008	1	1	0
Oxygen	O	8	16	8	8	8
Boron	B	5	10.81	5	5	6
Gold	Au	79	197	79	79	118
Silver	Ag	47	107.9	47	47	61
Uranium	U	92	238	92	92	146
Potassium	K	19	39.1	19	19	20
Chlorine	Cl	17	35.45	17	17	18
Neon	Ne	10	20.18	10	10	10
Einsteinium	Es	99	252	99	99	153

Properties of Matter — Quizzes Answer Keys

Quiz 1. Experimental Science
Lessons 1–4

Number the steps of the scientific method in the correct order.

A. _2_ Ask a question.
B. _1_ Learn about something/Make observations.
C. _6_ Share your results.
D. _4_ Design a test and perform it.
E. _3_ Make a hypothesis.
F. _5_ Check your results/Is your hypothesis right?

Mark each statement as either True or False.

1. _F_ You must always have a correct hypothesis.
2. _T_ It is important to control variables in your experiments.
3. _F_ Qualitative observations always use numbers.
4. _T_ Quantitative observations can be more useful to scientists than qualitative observations.
5. _T_ It is usually easier to make conversions between units in the metric system than in the Old English/American system.
6. _T_ A millimeter is smaller than a meter.
7. _F_ A graduated cylinder should be used to measure mass.
8. _T_ God has established laws to govern how chemicals react with each other.
9. _F_ Science can always tell us why things happen.
10. _F_ Matter has no mass.

Short answer:

11. Describe what chemistry is the study of. **Chemistry is the study of matter and how it reacts.**

Challenge questions
Short answer:

1. Is the measurement of the intensity of light from a distant star origins science or observation science? **Observation.**
2. Is the use of distant starlight to date the universe an example of origins science or observation science? **Origins.**

3. Why shouldn't you look through the eyepiece while lowering the objective on a microscope? **You could run the lens into the slide, causing damage.**
4. How are microscopes similar to telescopes? **They both use lenses to make an image larger.**
5. How are microscopes different from telescopes? **Microscopes only use lenses; telescopes sometimes use mirrors, too; microscopes are used to view tiny objects; telescopes are used to view far away objects.**

Match the scale with what phenomenon it describes.

6. _C_ Mohs scale
7. _A_ Fujita scale
8. _E_ Saffir-Simpson scale
9. _B_ Beaufort scale
10. _D_ Richter scale
11. _F_ Mercalli scale

Quiz 2. Measuring Matter
Lessons 5–9

Match the term with its definition.

1. _C_ The amount of a substance.
2. _H_ How strongly something is pulled on by gravity.
3. _A_ Matter cannot be created or destroyed.
4. _D_ How much space matter occupies.
5. _B_ How much mass is in a particular volume.
6. _G_ The ability for one substance to float in another.
7. _F_ Used to measure mass.
8. _E_ Used to measure weight.
9. _J_ A material that is denser than lead.
10. _I_ Only common material to become less dense when frozen.

Short answer:

11. Explain how the water you drink today could be the same water a dinosaur drank thousands of years ago. **Water is recycled. After a dinosaur drank water, it exhaled some water into the atmosphere. That water has been recycled through the water cycle for thousands of years.**

12. Explain what happens to nitrogen in the soil and in plants that demonstrates conservation of mass. **Nitrogen is absorbed by plants, eaten by animals, then returned to the soil when the plant or animal dies. It does not get used up.**

13. If an object floats in one liquid but sinks in another, what does that tell you about the densities of the two liquids? **The first liquid is denser than the second.**

14. How would you determine the volume of a toy car? **Use the displacement method.**

15. How are buoyancy and density related? **One substance is buoyant in another if it is less dense than the other substance.**

Challenge questions

Short answer:

1. What are two units for measuring mass? **Grams, slugs.**

2. What are two units for measuring weight? **Newtons, pounds.**

3. If you perform an experiment and the mass of the resulting substance is less than the mass of what you started with, what is one likely explanation? **It is likely that a gas was produced and escaped from the experiment. A nuclear reaction converts a small amount of mass into energy.**

4. Which is likely to be more dense, a one-inch cube of steel or a one-inch cube of wood? **Steel is denser than wood.**

5. If you are traveling in a car with a helium balloon and the driver suddenly puts on the brakes, what will happen to your body and what will happen to the balloon? **Your momentum will carry your body forward as the car suddenly slows down. It will also carry air molecules forward. Helium is lighter than the air, so the balloon will move backward to take the place of the air molecules.**

Quiz 3. States of Matter
Lessons 10–15

Use the terms from the list below to fill in the blanks.

1. The three states of matter are _solid_, _liquid_, and _gas_.

2. _Adding heat_ causes the molecules in matter to move more quickly.

3. _Removing heat_ causes the molecules in matter to move more slowly.

4. _Removing heat_ is required to change a gas into a liquid.

5. _Adding heat_ is required to change a solid into a liquid.

Write S beside the statement if it describes a property of a solid, L if it describes a liquid, and G if it describes a gas. Some statements describe more than one state of matter.

6. _S, L_ Molecules are close together.

7. _G_ Molecules are far apart.

8. _L, G_ It takes on the shape of its container.

9. _G_ Molecules move very quickly.

10. _L_ Molecules slide over each other.

11. _G_ Easily compressed.

12. _S_ Has a defined shape.

13. _S, L_ Has a defined volume.

14. _S_ Molecules only vibrate.

15. _S, L_ Not easily compressed.

Mark each statement as either True or False.

16. _T_ Thick liquids have a high viscosity.

17. _F_ As the temperature of a gas increases, its volume decreases.

18. _T_ As the pressure of a gas increases, its volume decreases.

19. _T_ A ball will usually bounce better on a warm day than on a cold one.

20. _F_ Molecules in a viscous liquid are not strongly attracted to each other.

21. _T_ There is a direct relationship between the temperature of a gas and its volume.

22. _T_ Crystals are more likely to form when a solid cools slowly.

Challenge questions

Mark each statement as either True or False.

1. _T_ Solid water is less dense than liquid water.

2. _T_ Glass can be classified as an amorphous solid.

3. _F_ Evaporation requires that a liquid be heated to the boiling point.

4. _F_ Diffusion occurs as molecules move from an area of lower concentration to an area of higher concentration.
5. _T_ Glass does not have a definite boiling point.
6. _F_ All solids are denser than their liquid form.
7. _F_ Evaporation is slower on windy days.
8. _T_ Increasing surface area increases evaporation rate.

Identify each of the following changes as a chemical change or a physical change.
9. _Physical_ Adding water to orange juice.
10. _Physical_ Shredding a piece of paper.
11. _Chemical_ Taking aspirin for a headache.
12. _Chemical_ Burning a candle.
13. _Chemical_ Shooting off fireworks.

Quiz 4. Classifying Matter
Lessons 16–21

Match the terms below with their correct definition or description.
1. _C_ A combination of two or more pure substances where each keeps its own properties—a new substance is *not* formed.
2. _J_ A liquid with air bubbles trapped in it.
3. _B_ A substance made when two or more elements combine chemically.
4. _G_ The process of heating a mixture to kill the bacteria in it.
5. _A_ A substance that cannot be broken down chemically.
6. _D_ A mixture where the substances are thoroughly mixed up.
7. _E_ A mixture where the substances are not evenly mixed up.
8. _H_ A nearly universal solvent.
9. _I_ The process of breaking up fat into tiny pieces that can remain suspended.
10. _F_ The ability of fat molecules to keep air molecules suspended.

Short answer:
11. Explain why whipped cream eventually melts into a pool of white liquid. **The fat molecules become unable to hold the gas, and it escapes.**
12. Give an example showing that a compound does not act like the elements that it is made from. **Accept any reasonable answer, such as liquid water does not act like oxygen gas or hydrogen gas.**
13. Explain why water is considered by many to be a nearly universal solvent. **Most substances will dissolve in water because of its unique shape.**
14. What elements are found in the compound CH_4? **Carbon and hydrogen (1 carbon and 4 hydrogen atoms).**

Challenge questions
1. What is a mineral? **Compound found in earth's crust.**
2. Name three groups of minerals. **Silicates, carbonates, halides, sulfides, phosphates, oxides.**
3. What is a native mineral? **Mineral containing only one element.**
4. List three ways to separate substances in a mixture. **Filtering, decantation, distillation, evaporation, chromatography, centrifuge.**
5. Briefly explain how milk is turned into cheese. **Acid is added to coagulate the milk, curds are separated from the whey, an enzyme is added to harden the curds, curds can be pressed and aged.**
6. What instrument is used to separate blood cells from plasma? **Centrifuge.**

Quiz 5. Solutions
Lessons 22–28

Mark each statement as either True or False.
1. _T_ All solutions are mixtures.
2. _F_ All mixtures are solutions.
3. _F_ A saturated solution can dissolve more solute.
4. _T_ True solutions do not settle out.
5. _F_ Milk is a true solution.
6. _T_ Temperature affects how fast substances dissolve.
7. _T_ Surface area affects how fast substances dissolve.
8. _F_ A dilute solution has a high amount of solute.
9. _T_ The boiling point of a solution is affected by concentration.

10. _T_ Cold liquids can suspend more gas than warmer liquids.

Short answer:

11. Describe why decreasing temperature decreases the solubility of a solid in a liquid. **Decreasing the temperature causes the solvent molecules to move more slowly, so they cannot keep as many solute molecules separated as they could when they were moving more quickly.**

12. Describe why increased pressure increases the solubility of a gas in a liquid. **Pressure pushes the molecules closer together, so the gas molecules cannot escape as easily.**

13. What is a precipitate? **It is a dissolved substance that comes out of the saturated solution.**

14. Why is salt added to ice when freezing ice cream? **Salt lowers the freezing point of the ice, allowing it to absorb more heat from the ice cream mixture, and thus making the cream freeze more quickly.**

15. What is likely to happen to a car without antifreeze in the radiator? **The water will boil more easily and could boil over if the temperatures get high. Also, water will freeze more easily and could freeze in the winter time.**

Challenge questions

Short answer:

1. If a substance easily dissolves in water, would you expect it to easily dissolve in oil? **It is not likely since like dissolves like, and water and oil are very different.**

2. Why does soap easily dissolve in both water and oil? **Soap is a unique molecule that is shaped similarly to water on one end and similarly to oil on the other end.**

3. Name one commercial application for an emulsion. **Foods such as mayonnaise, lotions, paints, espresso, photographic film.**

4. How does the concentration of salt affect the boiling point of water? **As salt concentration goes up, so does the boiling point.**

5. Why does salt affect the boiling point this way? **The salt molecules prevent the water molecules from reaching the surface and escaping into the air, so more energy is needed to bring the solution to a boil.**

6. How does temperature affect the density of seawater? **The colder the water, the closer together the molecules are, so the denser it will be. This is true as long as the water does not freeze.**

7. Why is hard water considered a problem? **Hard water can cause soap scum to form on clothes, dishes, and skin. It can cause scaly deposits to build up in pipes and water heaters.**

8. Explain why cake is considered an emulsion. **Oil molecules are suspended in a sugar water solution. Then eggs are added as an emulsifier to keep the fat molecules suspended.**

9. Why is water often called the universal solvent? **There are a wide variety of things that will dissolve in water at least a little bit.**

10. Explain why it is easy to float in the Dead Sea. **The Dead Sea has very high salinity — a lot of salt dissolved in it. This makes the water more dense, making it easier for the water to support you.**

Quiz 6. Food Chemistry
Lessons 29–33

Choose the best answer for each question.

1. _B_ Which is the most popular drink in the world?
2. _A_ What are 70% of all soft drinks sweetened with?
3. _D_ To guarantee that each soft drink tastes the same, what must a company do?
4. _D_ Which of the following is not used to make soft drinks?
5. _D_ What accounts for the perceived flavor of a food?
6. _B_ Which civilization is believed to be the first to enjoy chocolate?
7. _C_ What is the purpose of fermenting vanilla beans?
8. _B_ Which of the following is a natural flavor?
9. _A_ How do additives help preserve food?
10. _C_ Which of the following helps bread to be fluffy?

Challenge questions

Choose the best answer for each question or statement.

1. _D_ Which chemical is the name for table sugar?
2. _C_ Which chemical gives tomatoes their red color?
3. _A_ A calorimeter measures energy in food by _burning_ it.
4. _B_ Flavor is a combination of _taste and smell_.
5. _B_ Food additives usually are not used for _calories_.

Properties of Ecosystems — Quizzes Answer Keys

Quiz 1. Introduction to Ecosystems
Lessons 1–6

Match the term to its definition.
1. _N_ Decomposer
2. _D_ Biotic
3. _B_ Ecology
4. _L_ Omnivore
5. _E_ Abiotic
6. _F_ Ecosystem/biome
7. _K_ Carnivore
8. _C_ Biosphere
9. _G_ Flora
10. _I_ Niche
11. _H_ Fauna
12. _A_ Habitat
13. _J_ Herbivore
14. _M_ Scavenger

15. Draw a food chain with at least three levels. Label the role of each organism (producer, consumer, etc.). **Accept reasonable answers.**

16. Draw a food web with at least six organisms. **Accept reasonable answers.**

Describe each of the following relationships.

17. Mutualism: **Relationship in which both species benefit.**

18. Parasitism: **Relationship in which one species benefits and the other is harmed.**

Challenge questions
Short answer:

1. Why do animals generally not migrate from one biogeographic realm to another? **Biogeographical realms are separated by large natural barriers such as oceans, high mountains, or large deserts.**

2. Describe the niche of a butterfly. **As a larva, the butterfly eats plants, its droppings fertilize plants, and it uses the plants for shelter and for a place to make its chrysalis. Adult butterflies drink nectar and pollinate different plants. Butterflies provide food for birds and other animals and provide beauty for people to enjoy.**

3. What is carrying capacity? **Carrying capacity is the maximum population an area can support.**

4. How does the number of first order consumers in a given area compare to the number of second order consumers? **There must be significantly more first order consumers than second order consumers. A good rule of thumb is 10 to 1.**

5. What law makes the oxygen, water, and nitrogen cycles necessary? **Law of conservation of mass/matter.**

Quiz 2. Grasslands & Forests
Lessons 7–11

Mark each statement as either True or False.
1. _T_ A puddle of water could be considered an ecosystem.
2. _T_ The amount of sunlight hitting the earth is affected by the tilt of the earth.
3. _F_ Polar regions are near the equator.
4. _F_ It is uncommon to have a fire in a grassland.
5. _T_ Grazing animals are specially designed to eat grass.
6. _F_ Burrowing animals make it harder for grass to grow.
7. _T_ Trees are the dominant plants in a forest.
8. _F_ Temperate forests are located near the equator.
9. _T_ Arboreal animals spend most of their time in trees.
10. _F_ Epiphytes are animals that live on the forest floor.
11. _T_ Rainforests receive over 80 inches of rain each year.
12. _F_ Pampas grass is very short.

Short answer:

13. Give three different names for grassland. **Can include pampas, savannah, prairie, or steppe.**

14. List the six different layers of a forest. A. **Emergent layer** B. **Canopy** C. **Understory** D. **Shrub** E. **Herb** F. **Floor**

15. List two types of trees that you are likely to find in a deciduous forest. **Oak, elm, beech, etc.**

16. List two types of trees that you are likely to find in a coniferous forest. **Pine, fir, spruce, etc.**

17. Name three common products that originally came from the tropical rainforest. **Pineapples, oranges,**

bananas, lemons, eggplant, peppers, cocoa, chicle, latex, aspirin, quinine, etc.

Challenge questions

Short answer:

1. Describe how succession might take place in a forest that was destroyed by a wildfire. **Small, quick-growing plants such as grass and dandelions will grow first. This will provide food for small animals such as rabbits and prairie dogs. These animals will attract predators such as coyotes and hawks. As the ground has more cover, there will be enough moisture for shrubs to begin to grow. The shrubs will provide habitat for nesting birds, squirrels, and other animals. Tree seeds will germinate and begin to grow. Eventually, the trees will dominate the area again, reducing the sunlight to the floor and causing some of the smaller plants to die out. The trees will provide shelter and food for large animals such as deer and bears.**

2. Explain how God designed grazing animals to survive in a grassland ecosystem. **Grazing animals have specially designed stomachs that allow them to digest grass. Different animals eat different parts of the grass plant, allowing the plant to feed more than one type of animal.**

3. Explain the purpose of each of the following parts of a tree.
    a. Outer bark: **Provides protection from hazards and harsh weather.**
    b. Phloem/inner bark: **Transports food from the leaves to the rest of the plant.**
    c. Cambium: **Generates new phloem and xylem cells.**
    d. Xylem/sapwood: **Transports water and nutrients from the roots to the leaves.**
    e. Heartwood: **Provides strength and structure.**

4. Place the following ecosystems in order from least amount of rainfall to greatest amount of rainfall. **Grasslands, coniferous forest, deciduous forest, tropical rainforest.**

5. Explain the importance of the tropical rainforests on the medical field. **One-fourth of all medicines are derived from plants that come from the rainforest.**

6. Describe what you might find in an ecotone between a grassland and a forest. **There might be more trees than in a grassland but fewer trees than in a forest. There is likely to be a combination of grassland and forest animals.**

Quiz 3. Aquatic Ecosystems
Lessons 12–17

Fill in the blank with the correct term from below. Not all terms are used.

1. _Phytoplankton_ are microscopic aquatic organisms that perform photosynthesis.
2. Plants and animals that live on the ocean floor are called _benthos_.
3. An _atoll_ is a coral reef formed around a sunken volcano.
4. _Nekton_ are animals that freely move throughout the ocean.
5. Where the ocean meets the land is called a _beach_.
6. An ecosystem where fresh water flows into the ocean is called an _estuary_.
7. Sudden rapid growth of algae is called an _algae bloom_.
8. Land along the banks of a river or stream is the _riparian zone_.
9. A _tributary_ is a smaller stream or river that flows into a larger stream or river.
10. The _inter-tidal zone_ is the part of the shore that is covered with water at high tide and uncovered at low tide.
11. A lake that is too shallow to have an aphotic zone is referred to as a _pond_.
12. A coral reef attached to land is a _fringing reef_.
13. _Plankton_ are plants and animals that move with the ocean currents.
14. The layer of water that sunlight is able to penetrate is the _sunlit/euphotic zone_.
15. _Overturn_ is the rapid exchange of cold and warm-water regions within a lake.

Short answer:

16. Briefly explain why you can expect to find more varieties of plants on a rocky beach than on a sandy beach. **Rocky beaches provide more cracks and soil for plants to anchor to.**
17. Why do coral grow only in relatively shallow water? **Coral rely on algae in their tissues to produce food for them. This requires sunlight for photosynthesis, so coral only grow where there is abundant sunlight.**

18. What is overturn in a lake? **Overturn is the rapid movement of cold water layers to the bottom of a lake and warm layers of water to the top of a lake.**

19. Why is an estuary a very productive ecosystem? **The water currents stir up and bring in a large amount of nutrients that spur plant growth.**

20. Which organisms form the base of the food chain in the ocean? **Plankton, particularly phytoplankton, produce most of the food that forms the base of the food chains in the ocean.**

Challenge questions

Mark each statement as either True or False.

1. _T_ Bioluminescent creatures produce light through a chemical reaction.
2. _F_ Coral bleaching occurs when there is too much bleach in the water.
3. _F_ Coral bleaching always causes the coral to die.
4. _T_ Algae and coral have a symbiotic relationship.
5. _T_ A dune system is an example of ecological succession.
6. _F_ A maritime forest usually has large trees.
7. _T_ The grass in a dune system helps to stabilize the dunes.
8. _T_ Dune grass must be tolerant to salt and wind.
9. _F_ Land can only be part of a single watershed.
10. _T_ The Mississippi River Basin is the largest watershed in the United States.
11. _F_ Water from the Mississippi River mixes quickly with the Gulf of Mexico.
12. _T_ The Great Lakes provide water and work for over 35 million people.
13. _F_ Invasive species are not a real threat to animals in the Great Lakes.
14. _T_ The Great Lakes can generate their own weather systems.
15. _F_ The Mississippi River has the largest volume of any river.
16. _T_ The Nile River is one of the longest rivers in the world.
17. _T_ River ecosystems vary as the speed of the river changes.
18. _T_ The Volga River is an important ecosystem in Russia.
19. _T_ The Amazon River has the largest watershed in the world.
20. _F_ Bioluminescent creatures live primarily in the sunlit zone.

Quiz 4. Extreme Ecosystems
Lessons 18–23

1. Place the animals below in the ecosystem(s) you are likely to find them **(8 points for each completed ecosystem)**.

Tundra	Desert	Oasis	Mountain	Cave
Arctic fox	Toad	Toad	Arctic fox	Scorpion
Reindeer	Lizard	Lizard	Moose	Bat
Moose	Snake	Snake	Ground squirrel	Cricket
Canada goose	Scorpion	Scorpion	Snake	Crayfish
Ground squirrel	Camel	Camel	Bat	
		Bat	Mountain lion	
			Big horned sheep	

Choose the best answer for each statement or question.

2. _B_ What is the layer of permanently frozen ground in the tundra called?
3. _A_ Which do not help plants survive in the tundra?
4. _A_ On average, how much moisture does the tundra receive each year?
5. _A_ On average, how much moisture does a desert receive each year?
6. _C_ Which of the following helps animals survive in the desert?
7. _D_ Which are you not likely to find in an oasis?
8. _B_ You would expect the temperature in an oasis to be _____ than in the desert.
9. _D_ Which ecosystem would not likely be found on a mountain?
10. _A_ What is the point above which no trees will grow?
11. _C_ Which is likely to increase as you go up a mountain?

12. _A_ Which animal are you likely to find only in Australian chaparral?

13. _C_ Which condition does not contribute to fire in the chaparral?

14. _D_ Which animal is most important to cave ecosystems?

15. _C_ Which of the following does not describe an animal that visits or lives in a cave?

16. _B_ Which sense is least useful inside a cave?

Challenge questions

Short answer:

1. List three ways that polar bears are designed to live in the tundra. **Layer of blubber, two coats of hair, webbed feet, white color, good swimmers, sharp claws, papillae on pads of feet.**

2. The largest hot desert in the world is the _Sahara Desert_.

3. Where is this desert located? **In northern Africa.**

4. How has this desert changed since the time of the Genesis Flood? **It used to be much wetter and supported animals such as elephants.**

5. List three major products that come from deserts. **Oil/petroleum, gold, diamonds, uranium, nickel, aluminum, sodium nitrate, copper, solar energy.**

6. What is the tallest mountain in the Himalayas? **Mt. Everest.**

7. Name three animals found only in the Himalayas. **Snow leopard, clouded leopard, Bengal tiger, red panda.**

8. List three possible fire cues for seed germination. **Heat, smoke, charred wood, oxidation, acids.**

9. What method do insect-eating bats use to find their food? **Echolocation.**

10. List three different kinds of food eaten by different kind of bats. **Insects, fruit, fish, blood.**

Quiz 5. Animal Behaviors

Lessons 24–27

Mark each statement as either True or False.

1. _T_ Hibernation is a seasonal behavior for animals.

2. _F_ An animal's heartbeat is higher during hibernation.

3. _F_ Estivation occurs during the winter.

4. _T_ Butterflies often migrate hundreds of miles.

5. _T_ Migrating birds often fly in a V formation.

6. _F_ Trickery is most animals' first defense.

7. _F_ Most animals are defenseless against their enemies.

8. _T_ Camouflage is a good animal defense.

9. _T_ Prairie dogs can alert their colony to possible dangers.

10. _T_ Adaptation can be a physical characteristic or a behavior.

11. _F_ Natural selection does not really occur.

12. _T_ Animals can adapt because their DNA allows for great variety.

13. _F_ Man is needed to maintain a balance in most ecosystems.

14. _T_ When the predator population increases, the prey population decreases.

15. _T_ God designed territoriality as a way to control populations of animals.

Short answer:

16. Describe how territoriality helps control animal populations. **If a male cannot find an adequate open territory, he will not mate until one becomes available.**

17. Explain why bears do not truly hibernate. **Bears' metabolism does not significantly slow down, and a bear can be awakened during the winter, so it is not truly hibernating.**

18. Describe one type of animal defense. **Flight, trickery, fighting, etc.**

19. Explain how adaptations are a result of creation and not evolution. **Adaptations are primarily a result of the variety put into DNA at creation, not because of evolutionary mutations accumulating to make new features.**

20. What is the most likely trigger for seasonal behaviors? **Change in the number of daylight hours.**

Challenge questions

Match the term with its definition.

1. _B_ Plant defense against animals.

2. _F_ Plant defense against weather.
3. _I_ Species developing from a common ancestor.
4. _A_ Birds commonly used to support evolution.
5. _C_ Ability of a species to survive better than others.
6. _G_ Organisms with human-modified DNA.
7. _H_ Chemicals used to kill unwanted animals.
8. _E_ Chemical used to control malaria.
9. _J_ Display done to attract a mate.
10. _D_ Noise made by male elk to attract a mate.

Quiz 6. Ecology & Conservation
Lessons 28–33

Identify each of the following as either natural (N), man-made (M), both (B), or unknown (U) in origin.

1. _N_ Cooler weather due to ash in the atmosphere from a volcanic eruption
2. _N_ Smoke in the air from a fire started by a lightning strike
3. _B_ Acid rain
4. _B_ Species extinction
5. _B_ Habitat reduction
6. _M_ Invasive species introduction
7. _M_ Captive breeding
8. _B_ Air pollution
9. _B_ Water pollution
10. _N_ Greenhouse effect
11. _U_ Global warming
12. _N_ Buffering capacity
13. _M_ Deforestation (**Accept B if natural disasters such as hurricanes are included.**)
14. _M_ Plastic recycling
15. _M_ Overhunting

Fill in the blank with the correct term.

16. The three R's of conservation are _reduce_, _reuse_, and _recycle_.
17. Ranch animals must share _land_, _food_, and _water_ with native animals.
18. Species that are no longer alive are said to be _extinct_.
19. The zebra mussel is considered an _invasive_ species.
20. The most costly captive breeding program was to save the _California condor_.
21. Three main areas of pollution include _air_, _water_, and _land_.
22. _Acid rain_ is caused by sulfur dioxide and nitrogen oxides in the air.
23. Materials that naturally decompose are said to be _biodegradable_.
24. An area's ability to neutralize acid rain is called its _buffering capacity_.
25. One thing I can do to help the environment is _(accept reasonable answers)_.
26. List four ways that people impact the environment. **Accept reasonable answers. Ways mentioned in the book include: clearing the land for homes and farming, ranching, building, industries that put pollution in the air, industries use natural resources, recreational activities, etc.**

Challenge questions
Short answer:

1. Briefly describe a biblical view of ecology. **The earth belongs to God; man is His steward. We are to take care of and use the resources God has given us for our good and God's glory, without misusing them.**
2. List two government organizations that are committed to protecting the environment. **US Fish and Wildlife, EPA, US Park Service, United Nations FAO.**
3. Briefly explain how ozone depletion can occur. **CFCs rise in the atmosphere, and UV radiation breaks off chlorine atoms which react with ozone to break it apart.**
4. List two alternative sources of energy that could replace fossil fuels. **Hydroelectric, solar, nuclear, wind, hydrogen cars, electric cars.**
5. Why are fossil fuels considered non-renewable resources? **Fossil fuels are not being made in any large quantities today.**

6. Explain two possible problems with plastic recycling. **People do not put forth efforts to recycle, some plastics are difficult to recycle, and it is more costly to recycle than to make new resin.**

7. Describe your plan for how you will be a good steward of God's environment. **Answers will vary.**

Properties of Atoms & Molecules — Quizzes Answer Keys

Quiz 1. Atoms & Molecules
Lessons 1–4

Label the parts of a helium atom.

A. _Electron_

B. _Proton or neutron_

C. _Neutron or proton_

B and C together form the _nucleus_.

Match the term with its definition.

1. _D_ Anything that has mass and takes up space.
2. _E_ A positively charged particle in an atom.
3. _I_ A negatively charged particle in an atom.
4. _B_ A neutral particle in the nucleus.
5. _A_ Mass of a proton or neutron.
6. _J_ Compact center of the atom.
7. _G_ Two atoms of the same element connected together.
8. _F_ Part of matter that cannot be broken down chemically.
9. _H_ Number of protons an element has.
10. _C_ Two or more atoms chemically bonded.

Challenge questions

1. Refer to the periodic table of the elements to complete the following chart.

Element	Atomic #	Atomic mass	# of protons	# of electrons	Most common # of neutrons
Carbon	6	12.01	6	6	6
Aluminum	13	26.98	13	13	14
Tungsten	74	183.9	74	74	110

2. Based on the electron configurations for the following elements, which would not be likely to bond with any other elements? **Argon (Ar).**

3. What is an isotope? **An atom having the same number of protons but a different number of neutrons.**

4. What is a valence electron? **An electron in the outer shell of an atom.**

5. What is the electron configuration for silicon (Si)? **2, 8, 4.**

Quiz 2. Elements
Lessons 5–10

Short answer:

1. What do elements in a column of the periodic table have in common? **The same number of valence electrons.**

2. What do elements in a row of the periodic table have in common? **Electrons filling the same energy levels.**

3. Which column of elements is most stable? **VIIIA.**

4. Elements in which column are most likely to react with elements in column VIA? **IIA.**

5. Elements in which column are most likely to react with elements in column VIIA? **IA.**

Write metal, metalloid, or nonmetal to match the type of element to its characteristics.

6. _Metal_ Silvery luster
7. _Metal_ Ductile
8. _Metal_ Conducts electricity
9. _Nonmetal_ Does not conduct electricity
10. _Metal_ Solid at room temperature

11. _Nonmetal_ Not shiny
12. _Metalloid_ Somewhat malleable
13. _Nonmetal_ Most often a gas
14. _Metalloid_ Semiconductor
15. _Metal_ Malleable

Mark each statement as either True or False.

16. _T_ Hydrogen is very reactive.
17. _F_ Oxygen is lighter than hydrogen.
18. _F_ Hydrogen is the most common element on earth.
19. _T_ All elements are recycled—they are not destroyed.
20. _T_ Carbon forms organic compounds.

Challenge questions

Match the term to its definition.

1. _B_ Column of the periodic table.
2. _F_ Row of the periodic table.
3. _A_ Metals in column 1.
4. _D_ Metals in column 2.
5. _E_ Man-made elements.
6. _G_ Ball-shaped carbon molecule.
7. _H_ Thread-like cylinders of carbon atoms.
8. _C_ Technology that combines hydrogen and oxygen to produce electricity.
9. _J_ Atoms of the same element that link together in different ways.
10. _I_ Elements that do not occur naturally.

Quiz 3. Bonding
Lessons 11–16

For each characteristic below, write I if it describes an ionic substance, C for a covalent substance, and M for a metallic substance. Some characteristics have more than one answer.

1. _I_ Formed by elements with very different levels of electronegativity
2. _I,M_ High melting point
3. _C_ Electrons are shared between two or three atoms
4. _C,M_ Insoluble in water
5. _I_ Forms ions
6. _I_ Electrons are given up or pulled away
7. _C_ Does not conduct electricity
8. _M_ Sharing of electrons on a large scale
9. _I,M_ Conducts electricity
10. _C_ Flexible

Short answer:

11. How are crystals formed? **When a liquid slowly cools, the atoms may line up in specific patterns to form crystals.**
12. What is the smooth side of a crystal called? **A face.**
13. What process is necessary for ceramics to become strong? **Heating or firing.**
14. What is the common ingredient in all natural ceramics? **Clay.**
15. Name three traditional ceramics. **Pottery, brick, porcelain, glass.**

Challenge questions

Mark each statement as either True or False.

1. _T_ Ionic bonding occurs between elements with very different electronegativities.
2. _F_ Ions are electrically neutral.
3. _F_ Ionic bonds occur between nonmetals.
4. _T_ Sodium fluoride is an ionic compound.
5. _F_ Covalent compounds easily conduct electricity.
6. _T_ Covalent bonds occur between nonmetals.
7. _F_ Metallic materials easily dissolve in water.
8. _T_ Metallic bonds have free electrons.
9. _T_ Metallic bonds form between elements with similar low electronegativities.
10. _T_ Brass is an alloy of copper and zinc.
11. _F_ Steel is an alloy of copper and tin.
12. _F_ Hydrates usually feel wet.
13. _T_ Hydrates can help prevent the spread of fire.

14. _T_ Resorbable ceramics are absorbed into the body.
15. _F_ Inert ceramics react with the body.

Short answer:

16. What is the difference between a cation and an anion? **Cations are positively charged ions while anions are negatively charged.**
17. Heating will usually remove water from hydrates. What is this process called? **Dehydration.**

Quiz 4. Chemical Reactions
Lessons 17–20

Mark each statement as either True or False.

1. _F_ All chemical reactions are fast.
2. _T_ A catalyst speeds up a chemical reaction.
3. _T_ Endothermic reactions use up heat.
4. _F_ A fireworks explosion is an endothermic reaction.
5. _T_ The same number of atoms must appear on both sides of a chemical equation.
6. _T_ Chemical equations demonstrate the first law of thermodynamics.
7. _T_ Reactants are on the left side of a chemical equation.
8. _F_ Catalysts are used up in a chemical reaction.
9. _F_ Inhibitors speed up a chemical reaction.
10. _T_ Sometimes inhibitors are helpful.
11. _T_ Catalysts lower the energy required for a chemical reaction to occur.
12. _T_ Exothermic reactions release energy.
13. _F_ The product of an exothermic reaction is cooler than the reactants.
14. _F_ Chemical reactions are rare.
15. _T_ Heat can increase the reaction rate.

Short answer:

16. What happens in a composition reaction? **Elements combine to form a new molecule.**
17. What happens in a decomposition reaction? **A molecule breaks apart into its individual elements.**

Challenge questions

Identify each of the following reactions as composition (c), decomposition (d), single displacement (sd), or double displacement (dd).

1. _Double displacement_
   $H_2SO_4 + 2\ LiOH \longrightarrow Li_2SO_4 + 2\ H_2O$
2. _Composition_ $P_4 + 10\ Cl_2 \longrightarrow 4\ PCl_5$
3. _Decomposition_ $CO_2 \longrightarrow C + O_2$
4. _Single displacement_
   $2\ Na + 2\ H_2O \longrightarrow H_2 + 2\ NaOH$
5. _Double displacement_
   $AgNo_3 + HCl \longrightarrow AgCl + HNO_3$

Short answer:

6. List three ways to increase the reaction rate of a chemical reaction. **Add heat, increase surface area of reactants, increase concentration of reactants, add a catalyst, reduce activation energy.**
7. List two groups of catalysts. **Heterogeneous, homogeneous.**
8. Which type of catalyst is found in a catalytic converter? **Heterogeneous.**
9. Which type of catalyst will bond with a reactant? **Homogeneous.**
10. What is the name for the energy stored in chemical bonds? **Enthalpy.**
11. If the enthalpy of the products of a reaction is higher than the enthalpy of the reactants, is the reaction endothermic or exothermic? **Endothermic.**
12. What is the first law of thermodynamics? **Matter and energy cannot be created or destroyed; they can only change forms.**

Quiz 5. Acids & Bases
Lessons 21–24

Choose the best answer for each question.

1. _D_ Which is not a type of chemical analysis?
2. _A_ pH indicators can tell the strength of which type of compound?
3. _B_ What flower can indicate the pH of the soil by the color of its flowers?
4. _C_ Which of the following is not an acid?

5. _A_ Which of the following is not a base?
6. _A_ What is formed when an acid combines with a base?
7. _B_ Which is not a characteristic of acids?
8. _A_ Which is not a characteristic of bases?
9. _A_ Which acid is the most produced chemical in the United States?
10. _C_ What common product is made primarily from salts?

Challenge questions

Choose the best answer for each statement.
1. _C_ Electroplating is depositing a thin layer of metal on a **_conductor_**.
2. _A_ Titration allows you to calculate how many **_molecules_** of acid or base are in an unknown sample.
3. _B_ A proton donor is another name for a(n) **_acid_**.
4. _C_ A proton acceptor is another name for a(n) **_base_**.
5. _D_ You can identify the acid in a chemical equation because it loses a(n) **_hydrogen_** atom.

Quiz 6. Biochemistry
Lessons 25–28

Short answer:
1. Identify two chemical reactions that sustain life. **Photosynthesis, digestion/cellular respiration.**
2. Name three main chemical compounds in food. **Carbohydrates, proteins, fat.**
3. What type of catalyst increases the rate of digestion processes? **Enzymes.**
4. What is an animal called that eats dead animals? **Decomposer or scavenger.**
5. Name two types of decomposers. **Bacteria, fungi.**
6. Name an element that is recycled by decomposers. **Nitrogen.**
7. Name three ways to keep farmland productive. **Crop rotation, lying fallow, fertilizers.**

8. Who was the discoverer of penicillin? **Alexander Fleming.**
9. What is the most common solvent in the human body? **Water.**

Match the term with its definition.
10. _B_ Kills unwanted insects.
11. _D_ Kills unwanted plants.
12. _H_ Kills unwanted fungus.
13. _C_ Farming without man-made chemicals.
14. _A_ Growing plants without soil.
15. _E_ Medicine to kill bacteria.
16. _F_ Medicine to encourage natural defenses.
17. _G_ Some of these plants have natural medicinal value.

Challenge questions

Short answer:
1. What are two conditions that inhibit enzyme reaction rate? **Heat and decreased pH (increased acidity).**
2. What condition most promotes decomposition? **Darkness and/or warmth.**
3. What are two controversies surrounding organic farming? **Are organic foods healthier? Can organic farms produce as much as non-organic farms? Can organic farms really control pests? Are GMOs bad for us? Is organic farming better for the environment?**
4. Briefly explain how chemotherapy works to treat cancer. **The chemicals target reproducing cells and prevent them from completing reproduction, thus killing the cells. Cancer cells are quickly reproducing, so they are killed faster than other cells.**

Quiz 7. Applications of Chemistry
Lessons 29–33

Briefly explain how chemistry is used in the making of each of the following items.
1. Perfume: **Solvent extraction or steam distillation is used to extract the scent molecules from flowers. These are then combined with alcohol to form perfume.**

2. Rubber: **Sulfur is added to rubber/latex, and then the mixture is heated to form molecules that are strong and flexible.**

3. Plastic: **Long, flexible polymers are formed from petroleum and then heated and molded into plastic.**

4. Fireworks: **Energy is added to chemical compounds to excite the electrons through explosions. When electrons return to their normal levels, they release light. Chemistry is also used in the combustion reaction of the black powder.**

5. Rocket fuel: **Liquid hydrogen and oxygen are combined at a high temperature to produce the combustion reaction that provides the needed thrust for lifting a rocket.**

Mark each statement as either True or False.

6. _T_ Vulcanization makes rubber useful in most temperatures.
7. _F_ Rubber is made from cellulose.
8. _F_ A polymer is a very short molecule.
9. _T_ Today, synthetic rubber is more widely used than natural rubber.
10. _F_ Perfume smells the same in the bottle as on your skin.
11. _T_ Latex is a natural polymer.
12. _T_ Bakelite was the first useful plastic.
13. _T_ Plastic is an important product in American life.
14. _T_ Fireworks are different colors because of different chemical compounds used.
15. _F_ Recipes for fireworks are freely shared.
16. _F_ Kerosene and carbon dioxide are common rocket fuels today.
17. _T_ Newton's third law of motion is important in rocket design.
18. _T_ Combustion is a chemical reaction that produces large amounts of heat.
19. _T_ Flower-scented perfume is made from the oil in flower petals.
20. _F_ Plastic is made from latex.

Challenge questions

Mark each statement as either True or False.

1. _F_ Scents smell the same on every person.
2. _T_ Silk is a natural polymer.
3. _T_ A milk protein can be used as a glue.
4. _F_ Creating polymers is very difficult.
5. _F_ Flames are always the same color.
6. _T_ Sodium chloride burns with a yellow flame.
7. _T_ Hypergolic rocket fuel is not very common.
8. _T_ Solid rocket engines must use up all of their fuel once they are ignited.
9. _T_ Liquid rocket fuel is used in most space rockets.
10. _F_ It is harder to control the rate at which cryogenic fuel burns than the rate at which solid rocket fuel burns.

Properties of Matter — Final Exam Answer Keys

Lessons 1–34

Fill in the blank with the correct term from below.

1. How much of a substance you have is its **mass**.
2. How much space something occupies is its **volume**.
3. How much gravity pulls on a mass is its **weight**.
4. The three states of matter are **solid**, **liquid**, and **gas**.
5. When a liquid changes to a gas, it is called **evaporation**.
6. When a solid changes to a liquid, it is called **melting**.
7. When a liquid changes to a solid, it is called **freezing**.
8. When a gas changes to a liquid, it is called **condensation**.
9. When a solid changes directly to a gas, it is called **sublimation**.
10. The thickness of a liquid is called its **viscosity**.

Match the type of quantitative measurement with the proper tool.

11. **B** Volume of a liquid
12. **D** Mass
13. **C** Weight
14. **E** Temperature
15. **A** Volume of a cube

For each characteristic or statement, put E if it describes an element, C if it describes a compound, or M if it describes a mixture. Some statements have more than one answer.

16. Cannot be broken by ordinary chemical processes. **E**
17. Contains two or more kinds of atoms. **C, M**
18. Always has the same ratio of elements. **E, C**
19. Iron **E**
20. Water **C**
21. Helium **E**
22. Air **M**
23. Seawater **M**
24. Only 92 of these occur in nature. **E**
25. Almost all substances on earth are these. **M**

Identify each of the following changes as either a physical change (P) or a chemical change (C).

26. Burning of a candle. **C**
27. Rusting metal. **C**
28. Freezing of water. **P**
29. Crushing a graham cracker. **P**
30. Combining oxygen and hydrogen to make water. **C**

Identify each characteristic as describing either a gas, a liquid, or a solid. Some statements have more than one answer.

31. Molecules are far apart. **Gas**
32. Has a definite shape. **Solid**
33. Easily compressed. **Gas**
34. Takes on the shape of its container. **Liquid, gas**
35. Molecules are very close together. **Solid, liquid**

Short answer:

36. If a liquid is cooled, will it be able to dissolve more or fewer solids? **Fewer.**
37. If a soft drink is very bubbly-looking, is it more likely to be warm or cold? **Warm.**
38. What similar processes are required to produce the flavors of vanilla and chocolate? **Both require fermenting, aging, and drying.**
39. How can you tell if a solution is saturated? **No more solute will dissolve.**
40. How can you tell if a liquid mixture is a solution or a suspension? **If it is a suspension, particles will settle on the bottom; if it is a solution, there will be no settling.**

Challenge questions

Identify each statement as origins or operational science.

1. _Origins_ All birds had a common reptile ancestor.
2. _Operational_ Horses give birth to horses.

Short answer:

3. List three different scales used to measure different kinds of storms. **Beaufort—wind; Fujita—tornado; Saffir-Simpson—hurricane; Mohs—mineral hardness; Richter—earthquake insensity; Mercalli—damage done by an earthquake**

Mark each statement as either True or False.

4. _F_ An object is denser than another object if it has a greater volume.
5. _F_ Rubbing alcohol is buoyant in water.
6. _T_ Solid water is less dense than liquid water.
7. _T_ Dissolving salt in water is a physical change.
8. _T_ During diffusion, molecules move from an area of higher concentration to an area of lower concentration.
9. _F_ A native mineral has two kinds of elements in it.
10. _T_ Chromatography uses paper to separate substances in a mixture.
11. _T_ Enzymes are used to harden cheese curds.
12. _F_ A centrifuge uses evaporation to separate substances in a mixture.

Match each word with its definition.

13. _C_ Amount of salt in a solution.
14. _D_ Water containing calcium and magnesium.
15. _F_ Energy to raise 1g of water 1 degree C.
16. _E_ 1000 calories (kilocalorie).
17. _B_ Flavor enhancer.
18. _A_ Prevents reacting with oxygen.

Properties of Ecosystems — Final Exam Answer Keys

Lessons 1–34

Match the term with its definition.
1. _C_ Flow of energy from one organism to another.
2. _B_ Area of the earth containing life.
3. _D_ Organisms that produce food.
4. _E_ Non-living.
5. _A_ The environment in which an organism lives.
6. _G_ Ability of the soil to neutralize acid.
7. _H_ Top layer of a forest.
8. _I_ Living life primarily in trees.
9. _F_ Grassland of Europe and Asia.
10. _J_ Microscopic aquatic animals.
11. _O_ Where fresh water flows into the ocean.
12. _M_ Land along the banks of a river or stream.
13. _N_ Coral reef formed around a sunken volcano.
14. _L_ Deep sleep during the summer.
15. _K_ Cavern in a mountain or underground.

16. Draw a picture of a food chain with at least three links. **Answers will vary.**
17. Draw a picture of a food web with at least six different and interconnected organisms. **Answers will vary.**
18. Draw a picture of the water cycle. **See student book, page 160.**
19. Put the organisms below in the ecosystem in which they are most likely to be found (**3 points for each completed ecosystem**).

Tundra	Grassland	Rainforest	Ocean	Cave
Heather	Sage	Lemon tree	Plankton	Millipede
Ephemerals	Grass	Pineapple	Algae	Bat
Polar bear	Zebra	Cocoa tree	Shark	Blind fish
Caribou	Gazelle	Monkey	Jellyfish	
Ptarmigan	Coyote	Tree frog	Octopus	
	Bison	Toucan	Crab	
	Capybara			

20. List at least 3 characteristics of each of the following ecosystems.

Coral reef	Estuary	Deciduous forest	Desert	Mountain
Warm, clear water	Fresh and salt water meet	Deciduous trees	< 10 in. rain/yr	Different ecosystems with altitude
Near equator	Very productive	30–60 in. rain/yr	Loses more moisture than it gains	Has timberline and snow line
Great biodiversity	Salt tolerant plants and animals	Cold winter	Plants/animals conserve water	Found everywhere in the world, even underwater
Coral		Warm/wet summer	Easily floods	

Mark each statement as either True or False.
21. _T_ Many different animals exhibit seasonal behaviors.
22. _T_ Bears do not truly hibernate since they can wake up in the winter.
23. _F_ Monarch butterflies complete their migration in one generation.
24. _F_ Female birds are the ones that usually sing and defend their territory.
25. _T_ All animals were originally designed to eat plants.
26. _F_ Trickery is the first instinct most animals have for defending themselves.
27. _T_ Animals can use claws and teeth for defense.
28. _F_ Territoriality is the primary way that populations are controlled today.
29. _T_ Territoriality is the original way God designed population control.

30. _F_ Natural selection does not really occur.
31. _T_ Pollution can sometimes have natural causes.
32. _T_ Recycling is one way to help reduce man's impact on nature.
33. _F_ We should all panic about global warming.
34. _F_ Man is the only reason species become extinct.
35. _T_ People have made great progress in reducing acid rain.

Challenge questions

Fill in the blank with the correct term.

1. The maximum population an area can support is called its _carrying capacity_.
2. A _ecotone_ is a transitional area between two ecosystems.
3. _Pioneer_ plants are the first plants to move into an area after a natural disaster.
4. The final stable ecosystem of a succession is called the _climax_ ecosystem.
5. _Mutualism_ is a relationship between two organisms in which both are benefited.
6. The land drained by a particular body of water is called a _watershed_.
7. _Echolocation_ is the use of sound waves by bats to detect objects.
8. Behavior performed to attract a mate is called _animal courtship_.
9. _Adaptive radiation_ refers to several species that develop from a common ancestor.
10. Ozone protects the earth from _ultraviolet radiation_ or **solar radiation**_.
11. _CFCs/chlorofluorocarbons_ are the molecules that are believed to cause ozone destruction. (**HCFS is also an acceptable answer.**)
12. Plastics are made from molecules called _polymers_.
13. Resources that cannot be replaced are _non-renewable resources_.

Short answer:

14. Draw a population pyramid for an aquatic ecosystem. **Diagram should include algae and/or phytoplankton on the bottom, zooplankton on the second level, small fish and other small aquatic animals on the third level, and larger fish or larger aquatic animals on the top.**

15. Describe a possible succession for an area of pine forest after a forest fire. **Small flowering plants will grow first, followed by grass and small shrubs; eventually, larger shrubs will crowd out the smaller plants; finally, trees will crowd out most of the grass and smaller shrubs.**

16. Draw a cross section of a tree trunk. Label all parts. **See student book, page 175.**

17. List at least three commercial products from the desert. **Oil/petroleum, gold, diamonds, uranium, nickel, aluminum, sodium nitrate, copper, solar energy.**

18. Explain how some plants can germinate only when there is a fire. **Some seeds require heat, smoke, or charring before they will germinate, so a fire is necessary for germination.**

19. Describe one animal courtship ritual you found interesting. **Answers will vary.**

Atoms & Molecules — Final Exam Answer Keys

Lessons 1–34

For each pair of elements, write I if they are most likely to form an ionic bond, C for covalent bond, or M for metallic bond.

1. _I_ Na + Cl
2. _C_ H$_2$ + O
3. _C_ O + O
4. _I_ K + Br
5. _M_ Al + Al
6. _I_ Mg + O
7. _C_ C + O$_2$
8. _M_ Ag + Ag
9. _M_ Cu + Cu

Note: elements on opposite sides of the periodic table are likely to form ionic bonds; elements that are both metals (from the left side) will form metallic bonds; elements from the right side (nonmetals) usually form covalent bonds.

Fill in the blanks with the terms from below.

10. A _catalyst_ can be used to speed up a chemical reaction.
11. The products of an _exothermic_ reaction have a higher temperature than the reactants.
12. The products of an _endothermic_ reaction have a lower temperature than the reactants.
13. An _enzyme_ is a catalyst that increases the rate of digestion.
14. An acid and a base combine to form a _salt_.
15. A substance is an _acid_ if it releases H$^+$ ions when dissolved in water.
16. A substance is a _base_ if it releases OH$^-$ ions when dissolved in water.

- Draw and label a model of a helium atom, which has an atomic number of 2 and an atomic mass of 4.

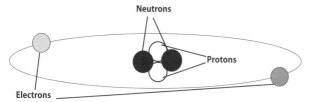

- Choose one of the following topics and briefly explain how chemistry plays a role in it: Farming; Medicine; The nitrogen cycle. **Farming: Nitrogen and other chemicals are used up in the growing of crops, so chemical fertilizers or other methods must be used to replace them. Also, insecticides, herbicides, and fungicides are all chemicals used to improve crop yield. Medicine: Chemicals are used to change the chemical reactions in the body to improve health. Nitrogen cycle: Nitrogen is used by plants, passed on to animals for their use, and then returned to the soil by decomposers.**

Match the term with its definition.

17. _B_ What natural rubber is made from.
18. _D_ What synthetic rubber is made from.
19. _E_ A long flexible chain of molecules.
20. _A_ Process that makes rubber strong and flexible.
21. _F_ A natural polymer found in plants.
22. _C_ Process of burning that releases large amounts of heat.

Short answer:

23. List three characteristics of a metal. **Silvery, solid, malleable, ductile, conduct electricity.**
24. List three characteristics of a nonmetal. **Not shiny, poor conductor, usually gas, brittle if solid.**
25. Explain the chemical reaction involved in your favorite experiment from this book. **Answers will vary.**

Challenge questions

1. Use the periodic table of the elements to complete the following chart.

Element	Symbol	Atomic #	Atomic mass	# electrons	# protons	Most likely # neutrons
Iron	Fe	26	55.85	26	26	30
Potassium	K	19	39.1	19	19	20
Mercury	Hg	80	200.5	80	80	120 or 121
Krypton	Kr	36	83.8	36	36	48

Fill in the blanks with the terms from below.

2. Temperature can increase the _**reaction rate**_ of a chemical reaction.
3. The elements in the first column of the periodic table are _**alkali metals**_.
4. The elements in the last column of the periodic table are _**noble gases**_.
5. The elements in the center of the periodic table are _**transition metals**_.
6. The elements in the second column of the periodic table are _**alkali-earth metals**_.
7. A bioceramic that does not react with the body is an _**inert ceramic**_.
8. A bioceramic that dissolves in the body is a **resorbable ceramic**_.
9. An acid is a _**proton donor**_.
10. A base is a _**proton acceptor**_.
11. Molecules that have water bonded to them are _**hydrates**_.
12. A _**homogeneous catalyst**_ is in the same phase as the reactants.
13. A _**heterogeneous catalyst**_ is in a different phase from the reactants.

Appendices

for Use with

God's Design: Chemistry & Ecology

Properties of Matter Master Supply List

The following table lists all the supplies used for *God's Design for Chemistry & Ecology: Properties of Matter* activities. You will need to look up the individual lessons in the student book to obtain the specific details for the individual activities (such as quantity, color, etc.). The letter *c* denotes that the lesson number refers to the challenge activity. Common supplies such as colored pencils, construction paper, markers, scissors, tape, etc., are not listed.

Supplies needed (see lessons for details)	Lesson
☐ Alum (in spice section)	28
☐ Baking soda	6c, 10, 17, 20, 22c, 24c, 25, 29, 34
☐ Balloon (helium-filled, optional)	9c
☐ Balloons (latex)	2, 6c, 15, 35
☐ Battery (6-volt)	17
☐ Bean seeds	18c
☐ Block (wooden)	12
☐ Bottle (plastic ½-gallon or 1-liter)	15c, 20
☐ Box (small)	3, 7
☐ Bread	30, 33c
☐ Cake mix	23c
☐ Candle	20, 35
☐ Charcoal briquettes	28
☐ Cinnamon	25
☐ Club soda	25
☐ Coffee filter	19
☐ Corn syrup	25
☐ Cornstarch	34
☐ Cotton balls	28
☐ Cups (clear)	3, 9, 22, 22c, 24, 26c, 28
☐ Cups (paper)	5, 6
☐ Dish soap	13, 15c, 22c
☐ Eggs	23, 23c, 27
☐ Eraser	8c

Supplies needed (see lessons for details)	Lesson
☐ Flour	30, 33
☐ Food coloring	25
☐ Funnel	19
☐ Goggles	28
☐ Golf ball	8
☐ Hammer	28
☐ Hand lotion	13
☐ Hand mirror	11
☐ Hole punch	5
☐ Honey	12, 13
☐ Ice tray	11
☐ Iodine	30, 34
☐ Jar (with lid)	11c, 17, 20, 21
☐ Jigsaw puzzle	16
☐ Lemon juice	10, 23, 25, 32
☐ Life Savers® candies (roll)	22, 31
☐ Marbles	8c, 9c
☐ Margarine	1, 33
☐ Meter stick/metric ruler	1, 3, 4, 5, 7
☐ Microscope and slides (optional)	3c
☐ Milk (not skim)	21c, 26, 31, 33
☐ Milk jug (1-gallon)	15
☐ Modeling clay	9, 9c
☐ Molasses	2
☐ Mustard (dry)	23
☐ Oil (spray)	33
☐ Oil (vegetable)	9, 13, 23, 22c, 30, 34
☐ Orange juice	19, 25
☐ Paper bag (brown)	30
☐ Paper clips	4, 5, 8
☐ Paprika	23
☐ Peanut butter	30
☐ Pennies	5, 8
☐ Perfume	14c

Supplies needed (see lessons for details)	Lesson
☐ Ping-pong ball	8
☐ Plastic bottles (empty 2-liter)	2, 6c, 18c, 28
☐ Plastic tub	9
☐ Plastic zipper bags	22, 26, 28, 33c
☐ Popcorn	9
☐ Potassium salt (in spice section)	22c, 24c
☐ Potato or tortilla chips	30
☐ Potting soil	18c
☐ Powdered sugar	34
☐ Pudding mix (instant)	31
☐ Rocks	12, 28
☐ Rolling pin	22
☐ Rubber band	5
☐ Rubbing alcohol	9, 34
☐ Salt	10, 22c, 23, 24c, 26, 26c, 27, 33
☐ Sand	28
☐ Scale (gram)	6c
☐ Silly Putty®	12
☐ Soft drink (canned, diet & regular)	24, 25c
☐ Soil	28
☐ Spices (ginger root, mint leaves, cinnamon sticks, all spice, cloves, peppermint oil, almond extract, etc.)	25

Supplies needed (see lessons for details)	Lesson
☐ Spoon (metal)	1, 6, 8c, 12
☐ Spoon (wooden)	1
☐ Stopwatch	1, 3, 26c
☐ Straw	27
☐ String	5, 18c
☐ Sugar	2, 21, 22c, 24c, 25, 26, 33
☐ Sugar cubes	6
☐ Tape (masking)	2, 3, 5
☐ Tape measure (cloth)	2, 15c
☐ Telescope (optional)	3c
☐ Tennis ball	3, 7c, 14
☐ Thermometer	2, 3, 26c
☐ Vanilla extract	21, 25, 26
☐ Vinegar	6c, 20, 21c, 23, 34
☐ Whipped cream (spray can)	21
☐ Whipping cream (liquid)	21
☐ Wire (copper)	17
☐ Wrenches (or other metal objects)	9
☐ Yeast	2, 33

Properties of Matter Resource Guide

Suggested Books

Structure of Matter by Mark Galan in the *Understanding Science and Nature* series from Time-Life Books—Lots of real-life applications of chemistry

Inventions and Inventors series from Grolier Educational—Many interesting articles

Molecules by Janice VanCleave—Fun activities

Chemistry for Every Kid by Janice VanCleave—More fun activities

Science Lab in a Supermarket by Bob Friedhoffer—Fun kitchen chemistry

Science and the Bible by Donald B. DeYoung—Great biblical applications of scientific ideas

200 Gooey, Slippery, Slimy, Weird & Fun Experiments by Janice VanCleave—More fun activities

Elements of Faith by Richard Duncan—Meaningful insights and spiritual applications from the periodic table of the elements

Chemistry by Dr. Dennis Englin—High school chemistry lessons and labs

Suggested Videos

Newton's Workshop by Moody Institute—Excellent Christian science series; several titles to choose from

Field Trip Ideas

- Visit the Creation Museum in Petersburg, Kentucky
- Visit a greenhouse or hydroponics operation to see the use of chemicals with plants
- Tour a battery store to learn about different types of batteries
- Visit a film processing plant to learn about chemicals in film processing or photo printing
- Visit a pharmacy
- Tour an injection molding plant to learn more about plastics
- Visit a farm to learn about the use of chemicals in farming

Creation Science Resources

Exploring the World Around You by Gary Parker—More detailed look at different aspects of ecology

Answers Book for Kids Vol. 1–8 by Ken Ham with Cindy Malott—Answers children's frequently asked questions

Creation: Facts of Life by Gary Parker—Good explanation of the evidence for creation

The Young Earth by John D. Morris, PhD—Evidence for a young earth

The New Answers Books 1–4 by Ken Ham and others—Answers frequently asked questions

Zoo Guide and *Aquarium Guide* by Answers in Genesis—A biblical look at animals, including extinction, defense/attack structures, biomes, and stewardship

Properties of Matter Works Cited

Ardley, Meil. *Making Metric Measurements*. New York: Franklin Watts, 1983.

"BHA and BHT." http://chemistry.about.com/library/weekly/aa082101a.htm.

Biddle, Verne. *Chemistry Precision and Design*. Pensacola: A Beka Book Ministry, 1986.

Brice, Raphaelle. *From Oil to Plastic*. New York: Young Discovery Library, 1985.

Busenberg, Bonnie. *Vanilla, Chocolate, and Strawberry: The Story of Your Favorite Flavors*. Minneapolis: Lerner Publications Co., 1994.

"The Chemistry of Cakes." http://www.margarine.org.uk/pg_app2.htm.

Chisholm, Jane and Mary Johnson. *Introduction to Chemistry*. London: Usborne Publishing, 1983.

Cobb, Vicki. *Chemically Active Experiments You Can Do at Home*. New York: J.B. Lippincott, 1985.

Cooper, Christopher. *Matter*. New York: Dorling Kindersley, 1992.

"Desalination of Water." *Columbia Encyclopedia*. 2000.

DeYoung, Donald B. *Science and the Bible*. Grand Rapids: Baker Books, 1994.

Dineen, Jacqueline. *Plastics*. Hillside: Enslow Publishers Inc., 1988.

Dunsheath, Percy. *Giants of Electricity*. New York: Thomas Y. Crowell Co., 1967.

"Energy Value of Food." http://www.cristina.prof.ufsc.br/digestorio/mcardle_energy_value_food_ch4_connection.pdf.

Erlbach, Arlene. *Soda Pop*. Minneapolis: Lerner Publications Co., 1994.

"Farming, Food and Biotechnology." *Inventions and Inventors*. 2000.

"Fleischmann's Yeast: Best-Ever Breads." Birmingham: Time Inc. Ventures Custom Publishing, 1993.

"The Flour Page." http://www.cookeryonline.com/Bread/flour.html.

Friedhoffer, Bob. *Science Lab in a Supermarket*. New York: Franklin Watts, 1998.

Galan, Mark. *Structure of Matter: Understanding Science and Nature*. Alexandria: Time-Life Books, 1992.

Groleau, Rick. "Buoyancy Brainteasers." http://www.pbs.org/wgbh/nova/lasalle/buoyancy.html.

"History." http://www.breadinfo.com/history.shtml.

"How and Why: Science in the Water." *World Book*. 1998.

"How Does a Water Softener Work?" http://home.howstuffworks.com/question99.htm.

"How Sweet It Is!" http://portal.acs.org/portal/fileFetch/C/CSTA_015104/pdf/CSTA_015104.pdf.

Hughey, Pat. *Scavengers and Decomposers: The Cleanup Crew*. New York: Atheneum, 1984.

Julicher, Kathleen. *Experiences in Chemistry*. Baytown: Castle Heights Press, 1997.

Kuklin, Susan. *Fireworks: the Science, the Art, and the Magic*. New York: Hyperion Books for Children, 1996.

Mebane, Robert C., and Thomas R. Rybolt. *Air and Other Gases*. New York: Twenty-first Century Books, 1995.

"Medicine and Health." *Inventions and Inventors*. 2000.

Morris, John D., Ph.D. *The Young Earth*. Colorado Springs: Master Books, 1992.

Newmark, Ann. *Chemistry*. New York: Dorling Kindersley, 1993.

Nottridge, Rhoda. *Additives*. Minneapolis: Carolrhoda Books Inc., 1993.

Parker, Gary. *Creation: Facts of Life*. Colorado Springs: Master Books, 1994.

Parker, Steve. *Look at Your Body—Digestion*. Brookfield: Copper Beech Books, 1996.

"Recipes Around the World." http://www.ivu.org/recipes.

Richards, Jon. *Chemicals and Reactions*. Brookfield: Copper Beech Books, 2000.

Saari, Peggy and Stephen Allison, Eds. *Scientists: The Lives and Works of 150 Scientists*. U.X.L An Imprint of Gale, 1996.

Solids, Liquids, and Gases. Ontario Science Center. Toronto: Kids Can Press, 1998.

Stancel, Colette, and Keith Graham. *Biology: God's Living Creation Field and Laboratory Manual*. Pensacola: A Beka Books, 1998.

Thomas, Peggy. *Medicines from Nature*. New York: Twenty-First Century Books, 1997.

VanCleave, Janice. *Chemistry for Every Kid*. New York: John Wiley and Sons, Inc., 1989.

VanCleave, Janice. *Molecules*. New York: John Wiley and Sons, Inc., 1993.

Walpole, Brenda. *Water*. Ada, OK: Garrett Educational Corp., 1990.

"William Prout." *Classic Encyclopedia*. http://www.1911encyclopedia.org/William_Prout.

Ziegler, Sandra. *A Visit to the Bakery*. Chicago: Children's Press, 1987.

Note: Some internet sites may no longer be available.

Properties of Ecosystems Master Supply List

The following table lists all the supplies used for *God's Design for Chemistry & Ecology: Properties of Ecosystems* activities. You will need to look up the individual lessons in the student book to obtain the specific details for the individual activities (such as quantity, color, etc.). The letter *c* denotes that the lesson number refers to the challenge activity. Common supplies such as colored pencils, construction paper, markers, scissors, tape, etc., are not listed.

Supplies needed (see lessons for details)	Lesson
☐ 1-gallon plastic zipper bag	32c
☐ 2-liter soda bottle	32c
☐ 3-ring binder	2
☐ Bag (produce)	19
☐ Baking soda	32c
☐ Box (small)	18, 23
☐ Cotton balls	18, 21
☐ Cups (clear)	15, 16
☐ Dividers (folder)	2
☐ Earthworms	2
☐ Eyedropper	15, 27
☐ Field guide to flowering plants	8
☐ Food coloring	12, 15
☐ Gloves (leather and cotton)	18
☐ Gloves (rubber)	30
☐ Goggles	14
☐ Grass and other plants	6, 8, 8c, 21, 23, 27, 31
☐ Hammer	14
☐ Ice	18
☐ Jar (with lid)	2, 6, 27, 32
☐ Leaves	20, 21
☐ Magnifying glass	1, 5, 14
☐ Meter stick/metric ruler	1, 8
☐ Microscope and slides (optional)	27
☐ Modeling clay	13, 30c

Supplies needed (see lessons for details)	Lesson
☐ Newspaper	8, 21, 30, 30c
☐ Oats	2
☐ Page protectors/sheet protectors	8, 30c
☐ Paint	21
☐ pH testing paper (optional)	27
☐ Photos of animals	18, 25
☐ Plastic zipper bags	14, 20
☐ Pots and pans	12, 27
☐ Potting soil	6
☐ Rocks	14
☐ Safety goggles	14
☐ Salt	15, 16
☐ Sand	2, 14
☐ Scale (bathroom)	30
☐ Seashells	14
☐ Soil	2, 6
☐ Spray bottles	31
☐ String	1
☐ Sunscreen lotion	30c
☐ Tagboard/card stock/poster board	7c, 8, 20c, 25, 32c
☐ Thermometer	16, 32
☐ Tissue paper or quilt batting	18
☐ Vinegar	31, 32c
☐ Water (distilled)	27
☐ World atlas	1c, 7, 17

Properties of Ecosystems Resource Guide

Suggested Books

Structure of Matter by Mark Galan in the *Understanding Science and Nature* series from Time-Life Books—Lots of real-life applications of chemistry

Inventions and Inventors series from Grolier Educational—Many interesting articles

Molecules by Janice VanCleave—Fun activities

Chemistry for Every Kid by Janice VanCleave—More fun activities

Science Lab in a Supermarket by Bob Friedhoffer—Fun kitchen chemistry

Science and the Bible by Donald B. DeYoung—Great biblical applications of scientific ideas

200 Gooey, Slippery, Slimy, Weird & Fun Experiments by Janice VanCleave—More fun activities

Elements of Faith by Richard Duncan— meaningful insights and spiritual applications from the periodic table of the elements

Suggested Videos

Newton's Workshop by Moody Institute—Excellent Christian science series; several titles to choose from

Field Trip Ideas

- Visit the Creation Museum in Petersburg, Kentucky
- Visit a greenhouse or hydroponics operation to see the use of chemicals with plants
- Tour a battery store to learn about different types of batteries
- Visit a film processing plant to learn about chemicals in film processing or photo printing
- Visit a pharmacy
- Tour an injection molding plant to learn more about plastics
- Visit a farm to learn about the use of chemicals in farming

Creation Science Resources

Exploring the World Around You by Gary Parker—More detailed look at different aspects of ecology

Answers Book for Kids Vol. 1–8 by Ken Ham with Cindy Malott and others—Answers children's frequently asked questions

Creation: Facts of Life by Gary Parker—Good explanation of the evidence for creation

The Young Earth by John D. Morris PhD—Evidence for a young earth

The New Answers Books 1–4 by Ken Ham and others—Answers frequently asked questions

Zoo Guide and *Aquarium Guide* by Answers in Genesis—A biblical look at animals, including extinction, defense/attack structures, biomes, and stewardship

The Ecology Book by Tom Hennigan and Jean Lightner—Multi-level, biology-focused title that unveils the intricate nature of God's world and the harmony that was broken by sin

Properties of Ecosystems Works Cited

"A Walk in the Forest." http://nationalzoo.si.edu/Education/ConservationCentral/walk/default.cfm.

"About the Dolphins." http://www.virtualexplorers.org/ARD/Dolphin/bkgd.htm.

"Alexander von Humboldt." http://geography.about.com/od/historyofgeography/a/vonhumboldt.htm.

"Alexander von Humboldt." http://www.humboldt.edu/~german/Alex.

"Alexander von Humboldt." http://www.phfawcettsweb.org/von.htm.

"Amazon River Animals." http://www.destination360.com/south-america/brazil/amazon-animals.php.

"Amazon." *Microsoft® Encarta® Online Encyclopedia*. http://encarta.msn.com/encyclopedia_761571466/Amazon_(river).html.

"Anaconda." *Microsoft® Encarta® Online Encyclopedia*. http://encarta.msn.com/encyclopedia_761553417/Anaconda_(snake).html.

"Animal Courtship Quiz." http://reference.aol.com/planet-earth/discovery/animal-courtship.

"Balance of Nature—Food Chains 101." http://www.hawkquest.org/TA/XL/Foodchain.pdf.

"Beach Exploratio." http://www.wetlandsinstitute.org/education/teacher/Beach_Exploration.pdf.

"Climate Change." http://www.epa.gov/climatechange/basicinfo.html.

"Coastal Ecosystems." http://www.soest.hawaii.edu/SEAGRANT/bmpm/coastal_ecosystems.html.

"Coral Reef Bleaching." http://www.marinebiology.org/coralbleaching.htm.

"Ecohysteria." http://www.apologiaonline.com/conf/ecohyst.pdf.

"Epiphytes." http://rainforests.mongabay.com/0405.htm.

"Facts About Endangered Species." http://www.endangeredspecie.com/Interesting_Facts.htm.

"Food Chains and Pyramids." http://www.mostateparks.com/onondaga/foodchain.htm.

Great Lakes Basin Ecosystem Team, US Fish and Wildlife Service. http://www.fws.gov/midwest/greatlakes.

"Great Lakes Fact Sheet." http://www.epa.gov/glnpo/factsheet.html.

"How Great is the Amazon River?" http://www.extremescience.com/AmazonRiver.htm.

"'Instant' Evolution Seen in Darwin's Finches, Study Says." http://news.nationalgeographic.com/news/2006/07/060714-evolution.html.

"Is Bleaching Coral's Way of Making the Best of a Bad Situation?" http://news.nationalgeographic.com/news/2001/07/0725_coralbleaching.html.

"Life in the Arctic Tundra." http://teacher.scholastic.com/products/instructor/Jan04_tundra.htm.

"Movie Review: Arctic Tale—Exaggerating the Effects of Global Warming." http://www.answersingenesis.org/articles/aid/v2/n1/arctic-tale.

"Ocean Science Activities." http://www.angelfire.com/la/kinderthemes/oscience.html.

Parker, Gary. *Exploring the World Around You*. Green Forest: Master Books, 2003.

"Peary, Robert E." http://www.pabook.libraries.psu.edu/palitmap/bios/Peary__Robert_Edwin.html.

"Recycling Plastics." http://www.eia.doe.gov/kids/energyfacts/saving/recycling/solidwaste/plastics.html.

"River Systems of the World." http://www.rev.net/~aloe/river.

"River." *Microsoft® Encarta® Online Encyclopedia*. http://encarta.msn.com/encyclopedia_761569915/river.html.

"Robert Peary." http://www.u-s-history.com/pages/h3896.html.

"The Emergence of Modern America." http://www.archives.gov/exhibits/american_originals/modern.html.

"The Estuary Ecosystem." http://www.teara.govt.nz/EarthSeaAndSky/MarineEnvironments/Estuaries/2/en.

"The Great River Amazon." http://hubpages.com/hub/THE-GREAT-RIVER-AMAZON.

"The Living Forest." http://www.arborday.org/trees/ringsLivingForest.cfm.

"The Tundra Biome." http://www.ucmp.berkeley.edu/exhibits/biomes/tundra.php.

"Theodore Roosevelt." http://edhelper.com/BiographyReadingComprehension_23_1.html.

"Theodore Roosevelt." http://www.desertusa.com/mag98/july/papr/du_troosev.html.

"Theodore Roosevelt." http://www.spiritus-temporis.com/theodore-roosevelt/presidency.html.

"Theodore Roosevelt." http://www.whitehouse.gov/history/presidents/tr26.html.

VanCleave, Janice. *Ecology for Every Kid*. New York: John Wiley & Sons, 1996.

Wilkes, Angela. *Usborne Book of Wild Places*. London: Usborne Publishing, 1990.

Note: Some internet sites may no longer be available.

Properties of Atoms & Molecules Master Supply List

The following table lists all the supplies used for *God's Design for Chemistry & Ecology: Properties of Atoms & Molecules* activities. You will need to look up the individual lessons in the student book to obtain the specific details for the individual activities (such as quantity, color, etc.). The letter *c* denotes that the lesson number refers to the challenge activity. Common supplies such as colored pencils, construction paper, markers, scissors, tape, etc., are not listed.

Supplies needed (see lessons for details)	Lesson
☐ Alka-Seltzer®	17c, 20c
☐ Ammonia (clear)	23, 23c
☐ Antacid tablets or liquid	23
☐ Baking soda	1, 12c, 17, 23, 24
☐ Balloons (latex)	30, 33
☐ Banana	26c
☐ Battery (9-volt)	12c
☐ Bible	35
☐ Borax	31c, 32c
☐ Bread	28
☐ Cabbage (red/purple)	21
☐ Candle	9, 10, 17
☐ Copper sulfate (available at swimming pool supply store)	32c
☐ Cornstarch	31c
☐ Cups (clear)	17c
☐ Cups (foam)	20c
☐ Cups (paper)	12c
☐ Diaper (disposable)	34
☐ Dish soap	10c, 23, 34
☐ Dry ice	10
☐ Eggs	7, 20
☐ Epsom salt	15, 32c
☐ Eyedropper	34

Supplies needed (see lessons for details)	Lesson
☐ Flashlight with battery	6
☐ Food coloring	34
☐ Garlic powder	28
☐ Gelatin	25c
☐ Geode (optional)	15
☐ Ginger ale	28
☐ Glitter	32
☐ Gloves (leather and cotton)	10
☐ Grass and other plants	27
☐ Hydrogen peroxide	19
☐ Jar (with lid)	17, 20, 22c, 29
☐ Lemon juice	19, 22
☐ Margarine	8, 28
☐ Marshmallows (mini, colored)	11, 12, 13
☐ Matches	9, 10, 17
☐ Mentos® candies	1c
☐ Milk (not skim)	22, 34
☐ Modeling clay	15c, 17
☐ Oil (olive)	12c
☐ Oil (vegetable)	8
☐ Paper clips	22c
☐ Paper towels	34
☐ Peanut butter	8
☐ Pennies	22c
☐ Pineapple juice (fresh, not frozen)	25c
☐ Pinecones	32c
☐ Plant food	27
☐ Plaster of Paris	15c
☐ Plastic zipper bags	26c, 34
☐ Plate (ceramic)	9
☐ Polymer clay (Fimo®, Sculpey®, etc.)	16
☐ Potassium salt (in spice section)	32c
☐ Potato	19
☐ Rubber band	30

Supplies needed (see lessons for details)	**Lesson**
☐ Rubbing alcohol	29
☐ Salt	12c, 15, 22c, 32c
☐ Silver object (tarnished)	14
☐ Silver polish/tarnish remover	14
☐ Soft drink (lemon lime)	22
☐ Soft drink (diet 2-liter bottle)	1c
☐ Spices (ginger root, mint leaves, cinnamon sticks, allspice, cloves, peppermint oil, almond extract, etc.)	29, 29c
☐ Starch (liquid)	34
☐ Steel wool without soap	10c, 20
☐ Stopwatch	17c, 20c
☐ Straw	33

Supplies needed (see lessons for details)	**Lesson**
☐ String	33
☐ Sugar	12c
☐ Swabs	24
☐ Tape (electrical or duct)	6, 33
☐ Test tubes	10c
☐ Thermometer	20
☐ Toothpaste (with fluoride)	7, 23
☐ Toothpicks	1c, 11, 12, 13
☐ Vinegar	1, 7, 17, 20, 22, 23c, 24, 25c
☐ Water (distilled)	12c, 23c
☐ Wire (copper)	6, 12c
☐ Yeast	26c

Properties of Atoms & Molecules Resource Guide

Suggested Books

Structure of Matter by Mark Galan in the *Understanding Science and Nature* series from Time-Life Books—Lots of real-life applications of chemistry

Inventions and Inventors series from Grolier Educational—Many interesting articles

Molecules by Janice VanCleave—Fun activities

Chemistry for Every Kid by Janice VanCleave—More fun activities

Science Lab in a Supermarket by Bob Friedhoffer—Fun kitchen chemistry

Science and the Bible by Donald B. DeYoung—Great biblical applications of scientific ideas

200 Gooey, Slippery, Slimy, Weird & Fun Experiments by Janice VanCleave—More fun activities

Elements of Faith by Richard Duncan—Meaningful insights and spiritual applications from the periodic table of the elements

Chemistry by Dr. Dennis Englin—High school chemistry lessons and labs

Suggested Videos

Newton's Workshop by Moody Institute—Excellent Christian science series; several titles to choose from

Field Trip Ideas

- Visit the Creation Museum in Petersburg, Kentucky
- Visit a greenhouse or hydroponics operation to see the use of chemicals with plants
- Tour a battery store to learn about different types of batteries
- Visit a film processing plant to learn about chemicals in film processing or photo printing
- Visit a pharmacy
- Tour an injection molding plant to learn more about plastics
- Visit a farm to learn about the use of chemicals in farming

Creation Science Resources

Exploring the World Around You by Gary Parker—More detailed look at different aspects of ecology

Answers Book for Kids Vol. 1–8 by Ken Ham with Cindy Malott and others—Answers children's frequently asked questions

Creation: Facts of Life by Gary Parker—Good explanation of the evidence for creation

The Young Earth by John D. Morris, PhD—Evidence for a young earth

The New Answers Books 1–4 by Ken Ham and others—Answers frequently asked questions

Zoo Guide and *Aquarium Guide* by Answers in Genesis—A biblical look at animals, including extinction, defense/attack structures, biomes, and stewardship

Properties of Atoms & Molecules Works Cited

"Alexander Fleming." http://www.pbs.org/wgbh/aso/databank/entries/bmflem.html.

"Bioceramics." http://www.azom.com/details.asp?ArticleID=1743.

Brice, Raphaelle. *From Oil to Plastic*. New York: Young Discovery Library, 1985.

"Buckyballs." http://scifun.chem.wisc.edu/chemweek/buckball/buckball.html.

"Charles Goodyear and the Strange Story of Rubber." *Reader's Digest*. Pleasantville, N.Y.: January 1958.

"Charles Martin Hall." http://www.geocities.com/bioelectrochemistry/hall.htm.

"Charles Martin Hall and the Electrolytic Process for Refining Aluminum." http://www.oberlin.edu/chem/history/cmharticle.html.

"Chemotherapy, What it is, How it Helps." http://www.cancer.org/docroot/ETO/content/ETO_1_2X_Chemotherapy_What_It_Is_How_It_Helps.asp.

Chisholm, Jane, and Mary Johnson. *Introduction to Chemistry*. London: Usborne Publishing, 1983.

Cobb, Vicki. *Chemically Active Experiments You Can Do at Home*. New York: J.B. Lippincott, 1985.

Cooper, Christopher. *Matter*. New York: Dorling Kindersley, 1992.

"Development of the Periodic Table." http://mooni.fccj.org/~ethall/period/period.htm.

"Diapers, the Inside Story." http://portal.acs.org/portal/fileFetch/C/CSTA_014946/pdf/CSTA_014946.pdf.

Dineen, Jacqueline. *Plastics*. Hillside: Enslow Publishers Inc., 1988.

Dunsheath, Percy. *Giants of Electricity*. New York: Thomas Y. Crowell Co., 1967.

"Enzyme Chemistry." http://www.math.unl.edu/%7Ejump/Center1/Labs/EnzymeChemistry.pdf?id=11897.

"Farming, Food and Biotechnology." *Inventions and Inventors*. 2000.

Galan, Mark. *Structure of Matter — Understanding Science and Nature*. Alexandria: Time-Life Books, 1992.

Helmenstine, Anne Marie, Ph.D. "Chemistry." http://chemistry.about.com.

"Historical Development of the Periodic Table." http://members.tripod.com/~EppE/historyp.htm.

"How and Why Science in the Water." *World Book*. 1998.

"How Does a Halogen Light Bulb Work?" http://home.howstuffworks.com/question151.htm.

Hughey, Pat. *Scavengers and Decomposers: The Cleanup Crew*. New York: Atheneum, 1984.

Julicher, Kathleen. *Experiences in Chemistry*. Baytown: Castle Heights Press, 1997.

Kuklin, Susan. *Fireworks: the Science, the Art, and the Magic*. New York: Hyperion Books for Children, 1996.

"Medicine and Health." *Inventions and Inventors*. 2000.

Morris, John D., Ph.D. *The Young Earth*. Green Forest: Master Books, 1998.

Newmark, Ann. *Chemistry*. New York: Dorling Kindersley, 1993.

Parker, Gary. *Creation Facts of Life*. Colorado Springs: Master Books, 1994.

Parker, Steve. *Look at Your Body: Digestion*. Brookfield: Copper Beech Books, 1996.

"Penny For Your Thoughts." http://www.tryscience.org/experiments/experiments_pennythoughts_athome.html.

Pinkerton, J.C. "Alexander Fleming and the Discovery of Penicillin." http://nh.essortment.com/alexanderflemin_rmkm.htm.

"Polymers: They're Everywhere." http://www.nationalgeographic.com/resources/ngo/education/plastics/nature.html.

Richards, Jon. *Chemicals and Reactions*. Brookfield: Copper Beech Books, 2000.

Saari, Peggy and Stephen Allison, Eds. *Scientists: The Lives and Works of 150 Scientists*. U.X.L An Imprint of Gale, 1996.

"Silly Putty." http://www.chem.umn.edu/outreach/Sillyputty.html.

Solids, Liquids, and Gases. Ontario Science Center. Toronto: Kids Can Press, 1998.

Student Activities in Basic Science for Christian Schools. Greenville: Bob Jones University Press, 1994.

Thomas, Peggy. *Medicines from Nature*. New York: Twenty-First Century Books, 1997.

VanCleave, Janice. *Chemistry for Every Kid*. New York: John Wiley and Sons, Inc., 1989.

VanCleave, Janice. *Molecules*. New York: John Wiley and Sons, Inc., 1993.

"Vulcanized Rubber." http://inventors.about.com/library/inventors/blrubber.htm.

"WebElements Periodic Table of the Elements." http://www.webelements.com/index.html.

Wile, Jay. *Exploring Creation with Chemistry*. Anderson: Apologia Educational Ministries, 2003.

Note: Some internet sites may no longer be available.

GOD'S DESIGN
FOR SCIENCE SERIES

EXPLORE GOD'S WORLD OF SCIENCE WITH THESE FUN CREATION-BASED SCIENCE COURSES

GOD'S DESIGN FOR LIFE
GRADES 3-8

Learn all about biology as students study the intricacies of life science through human anatomy, botany, and zoology.

GOD'S DESIGN FOR HEAVEN & EARTH
GRADES 3-8

Explore God's creation of the land and skies with geology, astronomy, and meteorology.

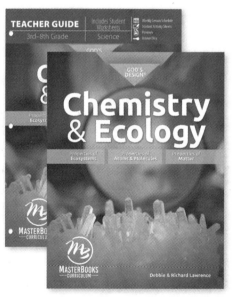

GOD'S DESIGN FOR CHEMISTRY & ECOLOGY
GRADES 3-8

Discover the exciting subjects of chemistry and ecology through studies of atoms, molecules, matter, and ecosystems.

GOD'S DESIGN FOR THE PHYSICAL WORLD
GRADES 3-8

Study introductory physics and the mechanisms of heat, machines, and technology with this accessible course.

AVAILABLE AT
MASTERBOOKS.COM 800.999.3777
& OTHER PLACES WHERE FINE BOOKS ARE SOLD

Daily Lesson Plans

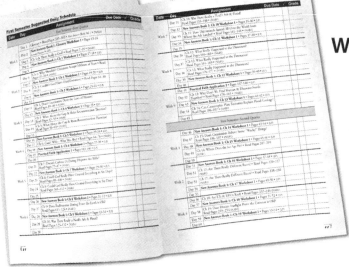

WE'VE DONE THE WORK FOR YOU!

PERFORATED & 3-HOLE PUNCHED
FLEXIBLE 180-DAY SCHEDULE
DAILY LIST OF ACTIVITIES
RECORD KEEPING

"THE TEACHER GUIDE MAKES THIN
SO MUCH EASIER AND TAKES T
GUESS WORK OUT OF IT FOR M

HOMESCHOOL

Master Books® Homeschool Curriculu

Faith-Building Books & Resources
Parent-Friendly Lesson Plans
Biblically-Based Worldview
Affordably Priced

Master Books® is the leading publisher of books and resources based upon a Biblical worldview that points to God as our Crea
Now the books you love, from the authors you trust like Ken Ham, Michael Farris, Tommy Mitchell, and many more are available as a homeschool curriculum.

MASTERBOOKS.c
Where Faith Grows.